Long distance ocean cruiser, 45
Custom designed to steel or other very ... methods. Accommodation fitted so that 1 or 2 can live for long periods at sea. Ketch rigged with essential sails and wind vane steering. Auxiliary engine and generator.

Gaff rigged cruising yacht, 40ft.
Elderly (built, say 1910) wood hull with most of original rig maintained as "vintage", but used for coastal cruising. Accommodation for 5 or 6, improved from year to year. Cutter rig. Auxiliary engine.

420 international two-man dinghy class 13ft. 9in.
High performance one-design glass fibre hull with centreboard. Crew of 2, one using trapeze. Sloop rig.

Laser international single-handed dinghy class, 13ft. 10·5in.
One-design glass fibre hull with mahogany centreboard. Single 76 square ft. sail.

Offshore trimaran, 50ft.
Custom designed for speed at sea. No ballast. Crew of 3 to 5. Accommodation in centre hull. Can race or cruise across an ocean. Sloop rigged, a few extra sails carried.

...ht, 60ft.
...igned to
...ula. Lead
...ith fully
...ace over
...umerous
...y engine.

The GUINNESS Book of

Yachting

Facts and Feats

Frontispiece
Twenty-fourth victor of the America's Cup, which represents world supremacy in yacht racing, the 12-metre class *Courageous*, designed by Sparkman and Stephens and owned by a syndicate of the New York Yacht Club. The cup has been retained by the club since 1851 against challengers from Britain, Canada and Australia. The America's Cup is the oldest international trophy in any sport.

12
US 26
26

The GUINNESS Book of Yachting Facts and Feats

Peter Johnson

GUINNESS SUPERLATIVES LIMITED
2 CECIL COURT, LONDON ROAD, ENFIELD, MIDDLESEX

By the same author

Passage Racing
Ocean Racing and Offshore Yachts
Yachtsman's Guide to the Rating Rule
Yachting World Handbook (editor)
Boating Britain

Published in Great Britain by
Guinness Superlatives Ltd., 2 Cecil Court,
London Road Enfield, Middlesex

ISBN 0 900424 30 3

Set in 'Monophoto' Baskerville Series 169,
printed and bound in Great Britain by
Jarrold and Sons Ltd, Norwich

Contents

'Talking of records is stupid, an insult to the sea. The thought of a competition is grotesque. . . .' Bernard Moitessier on arrival at Tahiti after being at sea longer and sailing more distance than any other single-handed sailor in his ketch *Joshua*, named after the first man to sail alone round the world.

Acknowledgements

Black and white illustrations are reproduced by kind permission of the following: Beken of Cowes 12, 13 (left), 48, 49 (both left), 68, 69, 82, 84, 99, 135, 137, 191, 192, 193, 197 (top right), 197 (bottom), 205, 206, 218 (bottom), 221, 232, 241, 242, 243, 244 (bottom), 245; Peter Johnson 13 (right), 40, 41 (left), 45, 46 (left), 51, 57 (right), 58, 86 (bottom right), 87, 92 (left), 93, 168, 200, 212 (bottom), 215, 225, 236, 238; Jonathan Eastland 14, 31, 49 (right), 56, 57 (left), 136 (bottom), 150, 178, 198, 223, 224; Barry Pickthall 67 (left), 85, 86 (top and lower left), 88, 89, 94 (bottom), 95, 176, 177, 247; Stanley Rosenfeld 43, 47, 59, 60, 63 (right), 81, 92 (right), 94 (top), 140, 141, 197 (top left), 207; Morris Rosenfeld and Sons 53, 62, 63 (left), 131, 133, 138, 185 (right); Yachting World 22, 25, 26, 66, 212 (top), 234; Foto-call 32, 33, 34, 35; Yacht Photo Service, Hamburg 46 (bottom and right), 96, 97; Bob Fisher 64, 65, 66; Hill's Studio 37, 181, 183; National Maritime Museum 36, 227, 244 (top); John Blomfield 40 (left), 50; Planet News 136 (top), 139; Huntsville Times 27; Mansell Collection 154; Syndication International 156; Herald-Sun 161 (bottom); Ed Lacey 106; Bill Robinson 41 (right), 42; Brian Manby 44; Universal Pictorial Press 52; The Press Association 125; Associated Press 8, 67 (right), 181.

Colour illustrations are reproduced by kind permission of the following: Beken of Cowes 38 (top right), 174 (bottom), 186 (top), 222 (top); Michael Barrington-Martin Studios 38 (top left), 55 (both top), 146, 222 (top); Alistair Black 19 (bottom), 110 (top right), 203 (top); Barry Pickthall 55 (bottom), 186 (bottom), 239; Yacht Photo Service, Hamburg 74 (top left), 91, 110 (top left); Jonathan Eastland 19 (both top), 175; Peter Johnson 74 (top right and bottom); Ken Hashimoto 38 (bottom); Bob Fisher 127; Brian Manby 222 (right and bottom); National Maritime Museum 110 (bottom); Stanley Rosenfeld frontispiece; Daily Telegraph Colour Library 203 (bottom).

Line drawings by Peter A. G. Milne.
Portrait drawings by Elizabeth Smith.

Introduction

Setting out on the water is always something of a venture. Man evolved on land, but there was some unknown *first* when he purposely used a floating object to get from one point on a waterway to another. Similarly for any individual there is always a first time afloat. It is a venture on such a first time and it always is again, if the craft is comparatively small and used for sailing, or rowing, or motor-boating.

Pleasure boating is often a long series of personal feats. After that first time out, can come the longest time out, the longest distance sailed, the biggest boat steered, the most rewarding trip – and the most exhausting one! The most congenial cruise may be remembered, while a single-handed sail is always an extreme case, being a trip by the smallest possible crew.

Travelling on water is particularly encompassed by numerous physical limitations (examined in appropriate chapters) and so it is a continual battleground for all types of competition and records. Such competition tends to require skill and ingenuity, but sometimes this is combined with sheer will power and at other times by the highest financial resources.

Racing – sail or power – is the obvious area for records and new achievements. Apart from personal ones, there are progressively larger fleets, faster boats, new and unusual classes, improved techniques, bigger boats, smaller boats. Yet cruising and voyaging are also peppered with records of distances, new routes, types of craft, age or sex of the sailor: new voyagers are forever looking for fields to conquer. The thousands who indulge all over the world in family and week-end cruising have their own share of feats in navigation and seamanship – sought after, or by accident of weather and lack of skill on some occasion.

The nomenclature of seagoing for pleasure is difficult in English. The French talk of 'navigation de plaisance', but 'pleasure sailing' has a somewhat degenerate ring which hardly fits its customary sporting and virile aspect. 'Yachting' has connotations of exclusive wealth, but this is a misuse and it is used here as an embracing term for sea, river and water activities in and around boats. It includes all types of non-commercial sail, power-boats for private use and racing power-boats. Pulling boats are used when yachting, but the sports of rowing, sculling and water-skiing are a specialist field. Ocean-rowing was so infrequent as to be classed as an individual feat, but in recent years there have been several remarkable voyages using oars as the only method of propulsion, so they are included here. Canoeing, both river and sea varieties, has also become established as a quite separate sport with Olympic status. Fishing makes use of boats, but is not boating.

Nineteenth-century yachting evolved from naval and commercial sailing-ships. The large yachts with big crews aped the sailing warships. The sailors were mainly naval reservists and sailing in private ownership kept them in training and discipline. This was the connection between the Royal Yacht Squadron and the Navy: today members of the Squadron – which is just a yacht club, though the most senior – still fly the Navy's White Ensign. The small yachts were copies of fishing-vessels, inshore sailing-tenders and work-boats. The change to purpose-built yachts was gradual and over a long period (one Bristol Channel Pilot Cutter was still used for ocean-racing in 1952), but the significance is that today's sailors for sport and leisure are the direct inheritors of sailing skills of many centuries. Yachting is the main receptacle for all the knowledge of sail which would otherwise have been lost. This it shares with other discontinued means of transport and war, which have become overwhelmingly sporting occupations: riding, archery, fencing, for instance.

One fascination of sail is, therefore, as the continued development of this historic means of propulsion – a means which was formerly the only communication between countries separated by seas and oceans. In feats concerning speed, distance, closeness to the wind and ability in bad weather, yachting continues to provide the arena. For this reason information of more general sailing and seagoing in earlier times is proper background here, though modern shipping is not part of the subject.

As the workday pressures of modern life grow more relentless, so escapist performances become more of a feature. Every year records of one sort or another are both attempted and claimed. This book of facts and feats may correct some misapplied attempts on that score. On the other side of the coin, some people will be encouraged to go even further, but be warned that the sea is not for stunts and has proved time and again that the foolhardy venture is doomed. The intention here is for neither of these uses, but rather that the world of boats is presented as enjoyable information and to give confidence and reassurance on the basis of the facts and the feats.

NOTES ABOUT DIMENSIONS

Miles whenever mentioned refer to nautical miles, unless otherwise stated. The nautical mile is 6080 feet (1853·2 metres), though the international nautical mile is officially 1852 metres but for practical purposes the same.

A *knot* is one nautical mile per hour and is equivalent to 1·16 statute miles per hour (m.p.h.).

A *cable* is one-tenth of a nautical mile and taken as about 200 yards.

A *fathom* is 6 feet.

When the *length* of a yacht is given, this always refers to the length over all (LOA), unless otherwise stated to be the length of the load waterline (LWL), designed waterline (DWL), rating (given in feet and tenths, or metres, depending on the class rules), or some other particular dimension. So '*Highland Light*, 68 ft 8 in' means that the yacht of the name has that length over all.

Measurement usage. As English is the predominant language of sailing, so English measure, particularly feet and pounds weight, are in use for description of yachts. Many classes too are known by a number ('the Seagull 28') which is its length in feet. This applies also to the Continent where the metric system is in everyday use, but boats are not infrequently given a class name in feet (metres cannot be approximated for length). Some classes are described specifically in metres as accepted class usage. In the parallel way, French and Italian sailors describe yacht ratings in feet. In the USA the signs are that feet and pounds will remain in use for many years. In a book on a subject, where the English-speaking countries have always and still do outnumber everyone else, it has been considered best usage to retain English measures.

1 Speed on the Water

UNDER SAIL

The slowest element on which to travel is water, but this is because of the great advances made in the speed of travel on land and the immense possibilities of air travel and space travel. In the nineteenth century, sailing speeds were not so disparate from those on land, while air travel did not exist.

The biggest vehicle for a thousand years was the sailing ship, until in the late nineteenth century sail gave way to steam, first for tugs and similar specialist steam craft, then for short-run commercial ships and then for the biggest vessels afloat. Sail remained longest for coastal trading craft and remains in this commercial role in several parts of the world. The sailing ship carried men, equipment and treasure from Europe to all parts of the world and back and had a permanent effect upon the politico-geographical pattern of the world as we know it today.

The unpredictability of sailing speeds was the principal factor in the demise of sail for international cargo-carrying and communication. Though ships could follow routes where winds were known to blow generally in a certain direction and be of a certain strength, this could not be relied upon. Early steamers were slower than the better sailing ships, but once steamers became reliable they could run to a closer timetable than a sailing ship ever could.

The first expansion of yachting occurred in early Victorian times at the end of the era of sail in commerce and war. In yachting the speed that mattered was one yacht against another or the ability to cruise and carry the owner and his guests in comfort. Sailing yachts of one sort or another thus became the successors, at least in the industrial and richer nations, of the commercial, naval, fishing and work boats which had sail as their means of propulsion. The result is that today with sailing techniques studied, practised and greatly improved, often with the help of materials that were not available before, the ability to sail to windward, or keep the sea under sail in all weathers, has not only been preserved but furthered. Some studies have even been made for renewed possibilities in commercial sailing ships because of the much greater efficiency that can now be obtained under sail. However, much of yacht performance, including actual speed, comes from the lightening of hulls and placing lead in the keel: a yacht has to move only itself and crew. Once a pay-load of any size has to be carried, many of the performance advantages would be lost. In waterborne commerce, therefore, development remains with screws or jets and not with spreading an area of cloth to the wind!

Sailing speeds are nearly all relative to some stipulated condition. In earlier days, these will reflect voyaging, passage-making and commercial considerations. Today, to the first two are added speeds over courses with the wind in a particular direction or records for certain regularly held races. The nearest thing to an absolute speed under sail would be a timed run for a very short period of time and, therefore, short distance. This would be without limit on wind direction and is the basis of a world sailing speed championship (see p. 23).

An accepted measure of speed in displacement vessels (as opposed to planing craft) is the ratio of velocity, V, to the square root of the load waterline, $\sqrt{L}$. The speed/length ratio is $V/\sqrt{L}$. This is an empirical ratio and the maximum achieved in conventional craft is usually in the region of 1·4. But this depends on the many facets of the design of any particular boat and the maximum may be lower or as high as 1·7. The value of the speed/length ratio is as a comparison within generic types of craft.

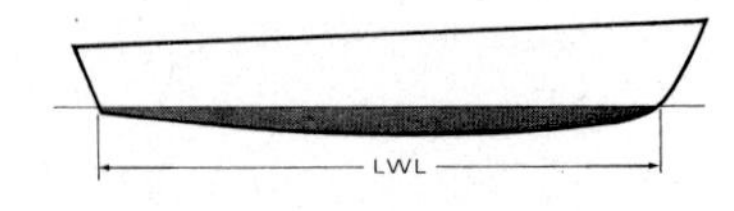

It is the length on the waterline (LWL) which limits the maximum speed of a conventional craft. Other things being equal, the longer the LWL the faster the potential speed.

Records of speed under sail before the end of the eighteenth century are vague because there was no system of measuring them. The time taken on a voyage would be known but is not of significance in determining the possible speed as the route taken would vary with wind direction. It is unlikely that more than 10 knots was ever achieved.

One of the first requirements for sailing speed came from smugglers so they could outsail revenue cutters. About 1790, shipowners in the United States became interested in fast cargo ships and America thereafter led in the world of commercial sailing-ship speed. Speeds were then encouraged and attempts were made to measure them. The use of the Yankee log and sand-glass meant errors up to 5 knots in speeds being measured. In heavy weather and big following seas when the highest speeds would be attained, it was the very worst conditions for taking celestial sights which could give accurate positions and thus the distance run for checking against time to give speed. Even after finding out the speed, the figure resulting was only a claim by the master or owners of the ship concerned. There was commercial advantage in being able to report high speeds: both for owners to obtain cargoes and for builders like the American, Donald Mackay, who specialized in fast clipper ships.

Sailing ships

The fastest speed over the period of a 4-hour watch recorded by a square-rigged sailing ship was 17½ knots by the five-masted *Preussen*. This was on a run in the South Pacific in 1903 under Captain B. Peterson. *Preussen* was the ultimate in big commercial square-riggers. She was 433 ft long, displaced 11 000 tons when laden and had a sail area of 60 000 ft^2, which developed 6000 horsepower. Her average speed in the trade winds was 8 knots. With a beam wind of 38 knots, when deep loaded with 8000 tons of cargo she sailed at 13¾ knots. *Preussen* was built in 1902 and run down in 1903 in the English Channel by the steamer *Brighton*. In November 1910, she was beached near Dover.

The five-masted *Preussen* was the ultimate in commercial square-rigged ships. Built in 1902 she sailed for only a few years.

The best day's run claimed for a square rigged sailing ship is 467 miles by the Black Ball liner *Champion of the Seas* on 11/12 December 1854. She was in the southern ocean on passage from Liverpool to Melbourne. The claim of an average of nearly 19½ knots is unlikely for the reasons given above. On the same voyage her average was only 199 miles per day; just under 8 knots.

The most credible best day's run for a sailing ship was that by *Lightning* under Captain Enright of 430 miles on 18/19 March 1857, a speed of just less than 18 knots average for the 24-hour period. *Lightning* was as fast as any square-rigger ever built. However the speed/length ratio for this run was 1·17. *Lightning* was built in 1854 by Donald Mackay of Boston, Mass., a famous builder of clipper ships. She was built for the Australia to England run along with three others of varying design (one of these was *Champion of the Seas*). Dimensions were 237½ ft on deck; 44 ft beam; 23 ft depth of hold and displacement 2084 tons. An example of her consistent speed is a 7-day recorded run of 2188 miles from 28 June to 4 July 1856: just under 13 knots.

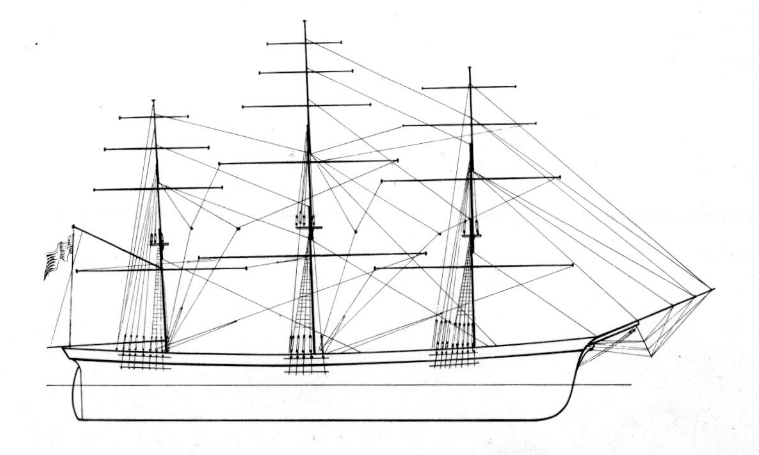

The fastest type of square-rigged commercial sailing ship was the clipper. This profile and spar plan of *Lightning*, built in Boston in 1854, is based on an LWL of 218 ft. A sustained speed of just under 18 knots was claimed.

Average speeds for the best clipper ships were more in the region of 6½ knots for an entire voyage from China or Australia to England, though the famous *Cutty Sark*, now preserved at Greenwich, averaged 8 knots. *Cutty Sark* was built by Scott and Linton and launched in 1869. Smaller than *Lightning* and much smaller than the later *Preussen*, she was 212½ ft on deck and 36 ft beam. Maximum recorded speed was 17½ knots, giving a speed/length ratio of 1·2. Such figures show the speeds attained by sailing-ships when they were of high commercial importance and for this reason make useful comparisons with sailing-boat speeds of later times.

Best times between England and Australia were 60 days, *Thermopylae*, 1868/9 and 1870/1, London to Melbourne; 63 days, *Lightning*, 1853, Melbourne to Liverpool. In 1875 *Cutty Sark* took 64 days on the outward passage from Lizard to Cape Otway. Transatlantic passages of note were 12 days 8 hours, *Adelaide*, 1864, New York to Liverpool; 14 days, *Howard D. Troop*, 1892, Greenock to New York.

The best sailing-ship time for a voyage round the world with a single stop was made by *James Baines* from Liverpool to Melbourne and back via Cape Horn in 133 days. The best time from London to Sydney, via the Cape and then Sydney to London via the Horn was by *Patriarch* in 1870 in 136 days, taking 67 days for the outward and 69 days for the homeward voyage. So for the total distance of 29 600 miles the average speed was 8·8 knots. (In 1973/4 *Great Britain II* skippered by Chay Blyth took 144 days in sailing time round the world from Portsmouth to Portsmouth but stopping off at Cape Town, Sydney and Rio.)

Yachts

Early yacht speeds would have been comparable with commercial and

From left to right
Britannia, 122 ft, designed by G. L. Watson, attained 12½ knots in a race in 1896.

Satanita, 131 ft 6 in displacing 126 tons with sail area of 10093 ft², sailed at 17 knots on a broad reach. She was built in 1893 to the design of J. M. Soper. To gauge her size note the men on deck.

Gipsy Moth V, 57 ft, was designed for Francis Chichester to attempt an average of 200 miles per day over 4000 miles.

naval ships of which the larger yachts were imitations and, therefore, speeds of yachts, as such, were not specially recorded. But when the era of big racing cutters began in the late nineteenth century, a few interesting speeds were recorded. In 1896 the cutter *Britannia* owned by the Prince of Wales attained 12½ knots on the last lap of a Channel race (speed/length ratio 1·3). Her waterline length was 86 ft 9 in. An even larger cutter, *Satanita* (waterline 93 ft 6 in), sailed at 17 knots on a broad reach in a strong wind in comparatively smooth water on the Clyde (speed/length ratio 1·5). Such a speed was only attainable because of her huge size, big sail area and the absence of very rough water. But the ratio is about the maximum expected for a conventional displacement hull.

Modern yachts also attempt speeds for a given voyage in the manner of the commercial sailing ships mentioned above. Some of these feats are mentioned in Section 4 in connection with ocean voyaging. But usually the voyage is for another primary purpose. For instance, Sir Francis Chichester, before his single-handed passage from Plymouth to Sydney, in *Gipsy Moth IV* decided to try and beat the 'average clipper time' of 100 days, but his main feat was in making the voyage single-handed non-stop which had never been done before. His time was 106 days.

This racing against a given time on one's own was not imitated by others. One more example of this private sport of Chichester was to attempt a single-handed voyage of an average of 200 miles per day over a distance of 4000 miles. For this a special yacht, *Gipsy Moth V*, was built (load waterline 41 ft 8 in). To find a suitable route for the required distance in the Atlantic, the voyage was made from Bissau in Portuguese Guinea to El Bluff in Nicaragua, Central America. In the attempt *Gipsy Moth V* actually sailed 179·1 miles per day average (7·46 knots). However, during the voyage

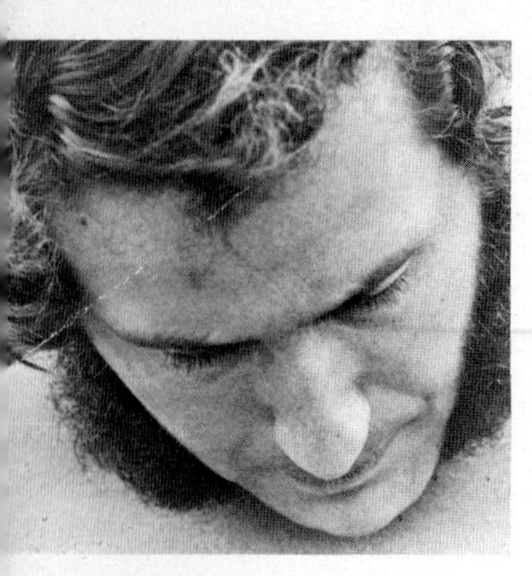

Alain Colas.

(January/February 1970), Chichester did achieve a run of 1018 miles in a consecutive 5-day period which was $203\frac{1}{2}$ miles per day (8·4 knots). Maximum sustained speed would, therefore, have been about 10 knots, and the yacht might have gone faster than this for very short periods of time, for instance when surfing on the front of a wave.

However, once even a self-created record attempt is made, then it is open to being broken. This was very soundly done by Alain Colas, the French single-hander, in the 67 ft trimaran *Manureva* (ex-*Pen Duick IV*). When sailing single-handed from Saint-Malo to Sydney and then back to Saint-Malo round the world, he achieved the 4000-mile run between two selected points on the chart in the Indian Ocean in 20 days – the 200 miles a day average. His best day's run was 326 miles. His average day's run for the nominal 29 600 miles of the whole trip was 175 miles. This was slightly less than Chichester which is surprising and shows the extent of the first achievement. It is also relevant to the type of yacht: a multihull generally puts up faster speeds in certain conditions, while a single-hulled displacement yacht has a narrower, slower speed range, but a more consistent one.

Term 'a fast yacht' is not very explanatory. Here white is faster than black in a light breeze, but when the wind begins to blow more strongly black's design of hull and sail plan makes her faster – at least on this point of sailing.

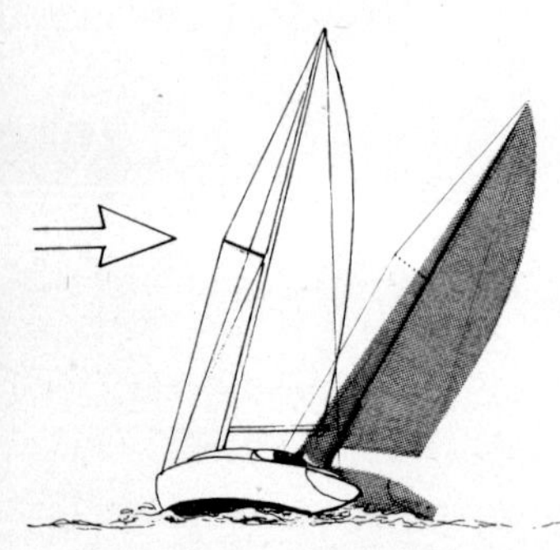

The stiffness of a yacht is an ability to remain as upright as possible against the heeling effect of wind.

Racing yachts receive the benefit of considerable research and development but not, paradoxically enough, towards absolute speed. They are designed to beat other racing yachts under specified rules. Many yachts are not even designed for fast maximum speed, but possibly for other types of performance. This might be light weather speed – for yacht A to go faster than yacht B in, say, 10 knots of wind. Yacht B might be faster in 25 knots of wind, but the expectation may be that such conditions prevail less often in the area where these yachts normally race. In this manner, sailing differs considerably from other speed sports. Another quality always required is windward ability to be able to outpoint rivals and sail closer to the wind; thus getting to a weather mark first. Off-wind speed is yet another quality and may be related to good steering characteristics. All such sorts of speed are highly comparative, but not in terms of knots.

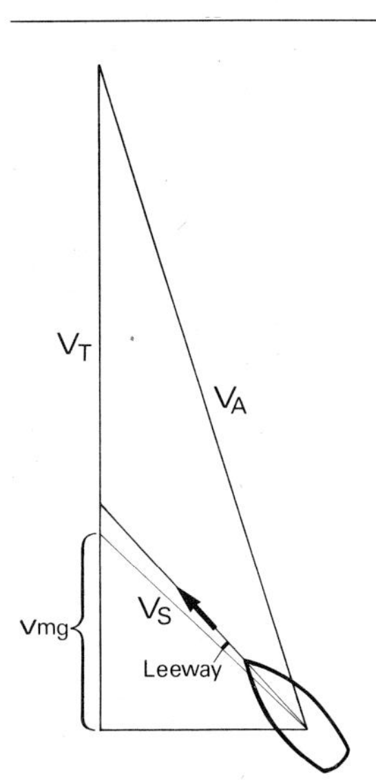

When making for a destination straight into the wind what really counts is the actual speed (or distance) made good to windward, Vmg. This diagram shows the components. The arrow is the heading of the yacht, but she is actually travelling at speed Vs because of leeway (side drift). VT represents the true wind blowing and VA is the apparent wind felt by those on board (and by the sails) because of the boat's own speed, Vs. For a sailing

Speed improvement in racing yachts is limited by class rules (see page 75). A simple example of this is sail area. Obviously more area of sail will mean higher speed but this is a fundamental means of restriction in all classes. Sail area will be limited by one set of rules or another, so improvements must come by the shape, type of cloth, range of controls or other variables related to the sails that *are* allowed.

A convincing win on an inshore course of, say, 6 miles could be three boat's lengths, but the speed difference between two boats this distance apart is less than one-tenth of a knot. The most likely reason for the result is a tactical one such as advantage gained in rounding a mark or by not running into a windless 'flat patch' or by tacking on a wind shift. A race is often a series of mistakes and the winner is the one who has made the least number of them, or the least serious errors.

Speed made good to windward is known as 'Vmg' and this depends on the speed through the water (Vs) and the angle to the wind. Some of the qualities to improve Vmg include the following. *Stiffness of the hull*, that is the ability to stand as upright as possible against the wind: this in turn is achieved by a low centre of gravity and a hull shape that resists heeling. The low centre of gravity comes from removing as much weight out of the hull (structure, accommodation and so on, though this is always restricted by rules and by practical limitations) and substituting it on the lead (or iron) keel. The hull shape aids stiffness when the centre of buoyancy of the heeled section moves out to give an effective lever arm. But as a hull shape has other requirements, this is usually partly sacrificed for other qualities (for example reduction of surface of hull to minimize friction). *Close-windedness* comes from the shape of any particular hull, a high-aspect ratio keel and rudder configuration helping to prevent leeway. A trim tab, as used on some racing boats, helps this ability. For any existing yacht these things are already designed in, but the shape of the sails set at any time has a marked effect. When sails become old and possibly shapeless and baggy the yacht sails less close to the wind.

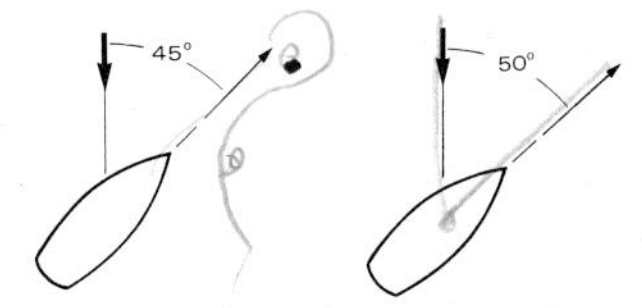

(Right) In this diagram both boats are going through the water at the same speed, but the right-hand one is not sailing so close to the wind and fails to make speed and distance to a windward destination.

Actual boat-speed when beating to windward is obviously a factor. (The term 'boat-speed' is used throughout sailing to mean the speed travelling along, as opposed to speed gained by tactics and smart handling.) Often this is called *footing* with the implication that the boat is allowed to travel fast at the expense of close-windedness. Sailing close to the wind at the expense of fast travel through the water yields less Vmg. But then fast travel, when the boat is allowed to pay off also lowers Vmg. Clearly there is an optimum as the diagram shows. Sailing too close not only can reduce boat speed, but will also increase leeway so the actual angle to the wind at which the boat is travelling is not even as good as the helmsman thinks in terms of the direction in which she is pointing. One further quality in getting to a windward point is *quick tacking* ability. This is apart from any tactics adopted, but if the boat can tack quickly it not only saves seconds up the course, but enables the helmsman to take advantage of changes in the wind without having to consider the time lost each time the yacht tacks. Quick tacking is the result of team work in the crew (whether 2 or 20), easily handled headsails with their ancillary gear, a short keel and good rudder control.

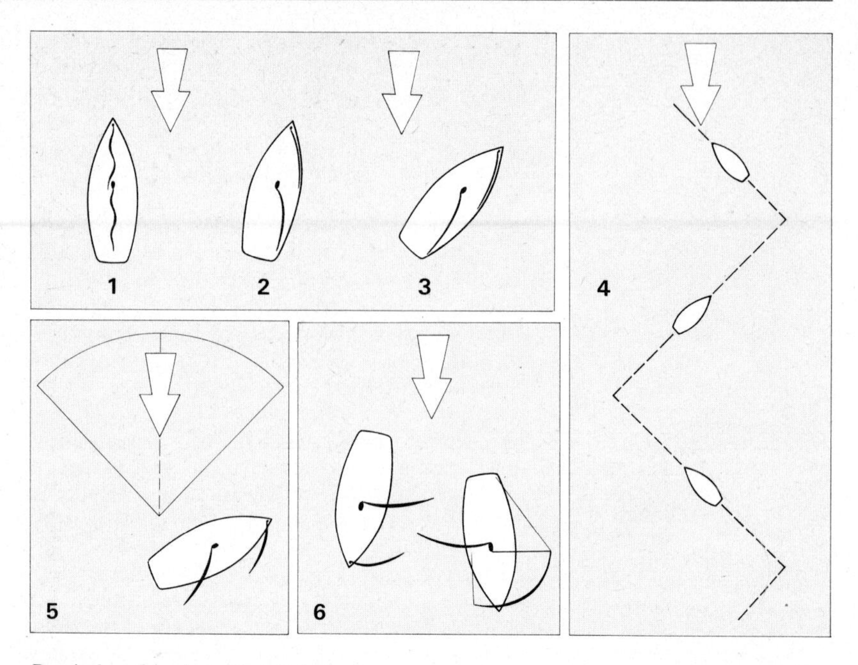

Top
Basic facts of sailing: no yacht can sail straight into the wind (1), but as she pays off (2) sails begin to fill, until at (3) she is sailing 'to windward'.

Far Right
Then a series of tacks (4) take her to a destination upwind.

Lower Left
Outside the 'forbidden zone' (5) sheets are eased off and the yacht 'reaches'.

Lower Middle
Running away from the wind is understood by all, but the side on which the sails are set (6) depends from which side of the stern the wind is blowing.

Reminder of basic sailing technique

Given a wind of more than 1 or 2 knots, any yacht or sailing-dinghy can sail in most directions, but not all. The boat cannot sail direct into the wind and if pointed direct into the wind the sail will merely shake like a flag. If pointed from either side of this windward direction at an angle of 45 degrees or less the sails will still flap and the boat will fail to make headway. But once at 45 degrees or more from the wind, provided the sails are pulled in hard, the yacht begins to move 'to windward'. This is the classic situation for a yacht and she is said to *beat* or *tack* to get to a destination upwind. In other words, a series of zigzags must be made, each leg of which is about 45 degrees to the direction from which the wind comes. The boat is said to be *close-hauled* when getting as close to the wind in this way as she can.

This beating into the wind can be long and tedious and if the current or tide is contrary, it may be impossible to make headway at all. Outside this 'forbidden zone' a boat will slide across the wind. In basic terms it might be said that the sails are pulled in until they just fail to flap and the boat is steered in the required direction. It is even easier to understand the basic technique when the boat is running away from the wind, as this is the point of sailing when it really is just being pushed directly by the force of the wind. Sails and rudder are adjusted to be able to steer in the precise direction. Though numerous developed techniques for improving performance on all points of sailing are now with us, the basic behaviour of any sailing boat in terms of wind direction is immutable.

Yet another quality of speed in a sailing boat is to be able to move quickly (or faster than another one if racing) when there is very little wind indeed. This is nothing to do with maximum potential speed of the hull.

Speed qualities

The simplest concept is sailing off the wind, as the boat does not make leeway and the course is set direct along the line of intended travel. Qualities for speed off the wind include the following.

Planing ability or *semi-planing* ability in keel boats enables the boat to exceed customary limitations on hull speed; roughly speaking it rides over its bow wave and gains dynamic lift, the speed dropping again suddenly as it 'comes off the plane'. *Light weight, flat sections* aft and *ample sail* area provide this effect. Sometimes emphasis on these is detrimental to windward abilities. For instance the shape of hull considered by a designer to increase downwind speed might adversely affect stiffness, one of the list of windward qualities.

Running away from the wind decreases the wind speed available, so extra sail area is needed; this is in the form of a spinnaker. *Spinnaker size* is, therefore, a major contribution to downwind speed. In classes where latitude is allowed in the design of spinnakers a decision can be made as to whether the spinnaker should play an important part in the performance. Allied with this is *spinnaker handling, hoisting* and *lowering*. A spinnaker up and drawing and then not allowed to collapse will ensure consistent speed off the wind. *Good steering* characteristics are a contribution to speed under spinnaker. The unbalancing effect of the extra sail area extended out from the boat can make some boats difficult to steer. Hull shape, keel and rudder configurations which alleviate this will have more speed. If steering difficulties do arise these are obviously in the higher wind speeds, from say 18 knots true wind upwards. In low wind speeds (7 knots true wind and below) downwind sailing becomes a problem because the boat 'running away' from the wind causes low apparent wind. *Ghosting* ability of a yacht can, therefore, give considerable speed advantage in such conditions. This is obtained from suitable *light weather sails* and frictional hull surface which is *very smooth* and *minimal* in area. Where there are centreboards and lifting rudders, these would be hauled up to accelerate this effect.

All these qualities give comparative speed between boats in a race whether they are of the same class, whether or not variations are allowed within the class (Section 2), or whether they are of quite different classes and types which show different advantages in speed in different conditions. Other things being equal a larger boat is faster than a smaller one. But things are invariably not equal and, for instance, a light centreboarded boat of 14 ft will frequently plane pass a heavy cruiser of 35 ft. In a class where design options are permitted, within rules of course, there may be a choice to design a boat to do best either on or off the wind. All such speed qualities are, it is emphasized, comparative and relative and regardless of handicap systems, if used, the object is to finish ahead of other yachts. Absolute speed is not sought.

Where there are record speeds for races, therefore, they will be for those courses that maintain the same geographical layout. The Prince of Wales Cup is the principal annual event for the International Fourteen Foot Dinghy class, but the venue and the exact length of course vary. Usually it is about 10 nautical miles, but an exact measurement is not necessary for a good race. When the wind is really strong and a 'record speed' might emerge, it is more than likely that the course will be shortened because of the bad weather. Where there are rules on the length of courses for a class they are likely to be in the form such as 'the windward leg is to be between 2½ and 3 miles'. So a class will not have a fastest time for its championship that is significant. Yacht races are not usually over a set distance like an athletic track event, nor for a set time like a team game in a stadium. There may be a time limit: if a boat is not in by a certain time, then the race does not count for her. Course records will apply only for the geographical course sailed annually or at other intervals between natural features or sea marks. An obvious example is the biennial Fastnet race starting at the Royal Yacht Squadron, Cowes, Isle of Wight, rounding the Fastnet Rock off south-west Ireland and finishing at the western end of Plymouth breakwater.

The J-class sloop *Ranger*, 135 ft, holds the record for the best time over a ten-mile course dead to windward.

Elapsed records

Record times for courses are elapsed times (the actual time taken by the first yacht on the assumption she crossed the start-line at the starting-gun) and not corrected times (Section 2). The reason is that corrected times depend upon the system of time allowance used and these vary from time to time and place to place.

Absolute speed

The highest authenticated speed made good to windward (Vmg) was by the American J class sloop *Ranger*, Harold S. Vanderbilt (waterline 87 ft). This was on 5 August 1937 in the final race of the America's Cup matches when she covered the 10-mile leg dead to windward in 1 hour 14 minutes 45 seconds, an average speed (Vmg) of 8·01 knots. The challenger *Endeavour II* (T. O. M. Sopwith) followed 4 minutes later.

The highest consistent speed for more than 100 miles recorded by the first-class racing cutters of the nineteenth-century was that made by *Britannia*, owned by the Prince of Wales, in her first season 1893. In a race for the Brenton Reef Cup organized by the Royal Yacht Squadron on 12 September *Britannia* covered 120 miles in 10 hours 37 minutes 35 seconds. The course was from a mark boat at the Needles, Isle of Wight, round the Cherbourg breakwater and back. On this reach both ways, she therefore averaged 11·3 knots, beating the other yacht, the American *Navahoe* by 57 seconds at the end. In the first 2 hours they logged 25 miles (12·5 knots). The yachts started at noon with double-reefed mainsails in a strong east wind. The time over the course, officially taken by the Squadron, is the fastest ever recorded for a race between the Solent and Cherbourg. *Navahoe* actually won the race on corrected time which was under American rules.

Interest in unconventional craft such as multihulls and hydrofoil boats in the early 1950s led to claims of high speed under sail by designers and manufacturers. To try and give an opportunity for settling some of these claims, a 'fastest boat' race was held in the Solent in 1954. This was on a four-sided course arranged without any beating. However the race was staged so that the course took over 4 hours. Inevitably stamina of crews and handling befogged the purer question of boat speed. The race was won by an Uffa Fox-designed, Fairey Marine-built Jollyboat class, 18 ft centreboarder sailed by Charles Currey. She narrowly beat *Ebb and Flo*, sailed and designed by T. Tothill. She was a catamaran with a twin athwartships rig.

In July 1954 at Cowes, a measured mile was set up on shore. Boats were invited to sail each way across this to attempt a sailing speed record. The Fairey Jollyboat sailed by Peter Scott did the fastest single run at 10·227 knots and a return run of 6·844 knots. Her winning average was declared at 8·535 knots.

In 1955 speed trials were again held off Cowes, but using the radar of HMS *Undine*. The advantage of this was that the direction to the wind would be optional for helmsman on any run, which had not been the case on the shore-based measured mile. The highest recorded speed was by the 18 ft catamaran *Endeavour* sailed by Ken Pearce at 14·6 knots.

The fastest speed claimed by a sailing boat is for the 30 ft hydrofoil-borne craft *Monitor*. The boat is said to have attained speeds close to 40 knots, but these are not substantiated. The boat was developed by J. G. Baker of

On the Cowes–Torquay–Cowes course in 1973, *Miss Embassy*. She was the first all-British-built gas-turbine offshore power-boat

Spirit of Ecstasy moving in power-boat race towards the chalk cliffs of the Isle of Wight

World sailing speed record attempt: a Unicorn class cat lifting on hydrofoils

RECORDS ON ESTABLISHED COURSES

Race	Distance (nautical miles)	Elapsed time	Date	Yacht	LOA	Owner	Speed (knots)
Newport–Bermuda	635	2d 20h 8m 22s	June 1974	*Ondine*	79 ft	Sumner A. Long (USA)	9·32
Fastnet race	605	3d 7h 11m 48s	August 1971	*American Eagle*	68 ft	Ted Turner (USA)	8·05
Channel race	225	25h 49m 13s	August 1973	*Sorcery*	61 ft 3 in	J. F. Baldwin (USA)	8·71
Cowes–Dinard (via Shambles L.V. Channel Islands to port)	173	21h 49m 13s	July 1973	*Apollo*	55 ft	Jack Rooklyn (Australia)	7·92
Plymouth–La Rochelle (west of Ushant)	355	50h 00m 28s	August 1971	*Bay Bea*	48 ft	Pat Haggerty (USA)	6·12
Bermuda–Plymouth	2870	14d 20h 15m 12s	July 1974	*Pen Duick VI*	74 ft	Eric Tabarly (France)	8·2
Middle Sea (from Malta, round Sicily, Pantellaria and islands)	613	89h 56m 08s	October 1973	*Aura*	49 ft	W. J. Stenhouse (USA)	6·8
Transpacific (Los Angeles–Honolulu)	2225	9d 5h 34m 22s	July 1971	*Windward Passage*	72 ft 9 in	Mark Johnson (USA)	10·02
Miami–Montego Bay (Jamaica)	811	3d 3h 40m	March 1971	*Windward Passage*	72 ft 9 in	Mark Johnson (USA)	10·72
Newport–Annapolis	473	2d 7h 40m 36s	June 1975	*Kialoha*	79 ft	John B. Kilroy (USA)	8·5
Sydney–Hobart	630	3d 1h 32m	December 1973	*Helsal*	72 ft 4 in	Tony Fisher (Australia)	8·56
Transatlantic (Sandy Hook–Lizard) (but fastest of any course)	2925	12d 4h 1m	1905	*Atlantic*	185 ft	Wilson Marshall (USA)	10·01
Transatlantic (single-handed) (Plymouth–Newport)	3000	21d 13h	1972	*Pen Duick IV*	70 ft	Alain Colas (France)	5·80
Round Britain (and Ireland)	2079	10d 4h 26m	1974	*British Oxygen*	70 ft	Robin Knox-Johnston	8·5
Round the World (Portsmouth via Cape Town, Sydney, Rio to Portsmouth)	27 120	144d 10h	1973–74 (Finished 11 April 1974)	*Great Britain II*	77 ft 2 in	Chay Blyth	7·82

Wisconsin, Illinois, for trials by the US Navy and as they formed part of a defence programme detailed data are not made public. In October 1956 (a time of high international political tension) *Monitor* was paced at 30·4 knots by a power-boat: presumably this had some form of water speedometer. The hydrofoil boat was reported to have sailed at twice the speed of the wind.

The most publicized high sailing speed in the USA in recent years has been that of the 32 ft D class catamaran *Beowulf V*. Owned by Steve Dashew with sail area of 408 ft^2 and weighing 880 lb, she was reported to have achieved 31·4 knots over an 815 metre course in a 20 knot breeze. The trials at which this speed was claimed were in summer 1974 and were run by the Pacific Multihull Association at San Pedro Bay, California. The timing was not, however, under any arrangements made by the national yachting authority or equivalent impartial body, nor have the timing systems been submitted to any independent scrutineering. Over such short runs the effect of timing on the ultimate speed figure is very critical.

No further trials were held in Britain until 1970. The lack of interest until then was partly due to the lack of a satisfactory way of accurate measurement afloat on what needed to be an optional course by the helmsman attempting the speed. In 1970, *Yachting World* organized some informal trials at Burnham-on-Crouch. New electronic measuring instruments designed for survey work were brought into use. In this experiment a Tornado class catamaran *Galadriel* sailed by Michael Day did 16·4 knots over a distance of 355 metres. In 1971 the trials were again held with observers of the Royal Yachting Association (RYA) present, but no higher speeds were achieved. The RYA then agreed to arrange official trials with prizes for the following year.

Conditions for a world sailing speed record were set up by the RYA and John Player Ltd. In 1972 £1000 was offered for the fastest yacht during a 10-day period on courses set in Portland Harbour; other boats within 5 knots of the winner were to share a prize of £1000. Conditions were that a sailing craft would enter on a 500 metre run, over which she would be timed by electronic survey equipment. A circular area with optional diameters along which the helmsman sailed was arranged. A run in only one direction was allowed. The yacht prior to, or during, the attempt had to accelerate from rest. Manual power only could be used on board. Substances could not be released from the hull to reduce surface friction.

Awards were available for attempts to be made elsewhere in the world during the year and all conditions applied together with the proviso that the attempt must be on water (not any other liquid nor on ice) and that the national yachting authority must supervise conditions and timings. In the event there were no other significant attempts other than at Portland.

The fastest boat under these conditions was *Crossbow* which sailed across the 500 metre course on 6 October 1972 at 26·3 knots. The wind was 19 knots at the time.

Crossbow was designed by Rod Macalpine-Downie specially for the speed attempt. As one run only was required, she was a one-sided proa with a 60 ft main hull and a 31 ft 6 in span with a pod in which the crew were situated to windward. Waterline when at rest, of the main hull was 50 ft and beam 1 ft 10 in. In half the wind speed at which the record was made the proa sailed at three-quarters of the record speed. The crew could walk from the main hull to the 15 ft pod as the wind increased. The yacht was

Left
The proa *Crossbow*, which after 1973 held the world sailing speed record over a 500 metre course.

Right
Icarus, a Tornado class catamaran on hydrofoils, travels at over 20 knots fully foil-borne.

Below
The profile and section of *Crossbow*, holder of the world sailing speed record over 500 metres. It could only sail on the starboard track.

unable to tack and at the completion of each run in the speed attempt, lowered sail and was taken in tow. *Crossbow* was owned by T. J. Coleman and T. Hall, who sailed the boat together with the designer and Reg White.

Second fastest yacht in 1972 was *Icarus*, a Tornado class catamaran on hydrofoils. The boat was run by a syndicate: J., B., and A. Grogono, D. Pelly and J. Fowler. The hydrofoils were a single set of light alloy foils with arcs of circular radius, sited one-third of the length of the hulls from the bows. One further pair of foils were on the lower edge of the rudder to prevent tip-forward capsize at high speeds. Her best speed was 21·6 knots in a 19 knot wind. Altogether 18 boats made speed attempts including a conventional Flying Dutchman Olympic centreboarder which, sailed by Rodney Pattisson, logged 10·7 knots.

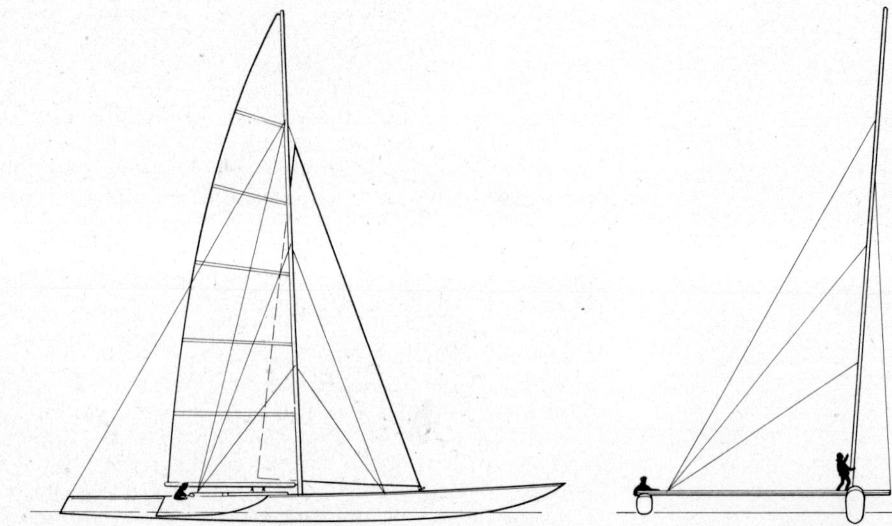

Sailing speed record

The fastest sailing boat speed in the world over a 500 metre run was found when the trials were repeated in October 1973. Under the same rules *Crossbow* made the fastest speed at 29·3 knots in a 20 knot wind. This is a speed/length ratio of 4·14. At this the world sailing speed record stands! *Crossbow*, under the same ownership, had been modified by shortening the over-all length of the hull to 55 ft. There was a new crew pod and this had a diamond-shaped hydrofoil under it to give it lift and prevent it touching the water. The height of the mainsail had been increased by 5 ft. On 30 September 1975, *Crossbow* beat her own record by attaining 31·09 knots.

The fastest speed by a production boat, which is already used by hundreds of owners was that of the Hobie Cat 14, which recorded 14·5 knots.

Future possibilities and limitations for speed attempts under sail have been shown by these trials. Despite the presence of a number of original types (multihulls, foilers, special rig craft and so on), the winner was by far the largest yacht. Higher speed from the wind would appear to be possible by an even larger boat: structural integrity and manual control are the limiting factors. It should also be noted that the prizes offered so far are only a fraction of the cost of preparing a large craft for a single 500 metre dash. Assuming sufficient area of protected water can be found to enable a large craft to have an optional course, then higher speeds might also be achieved in higher wind speeds, say 30 knots. But a proa, or any other configuration of vessel, would have a maximum hull speed above which no rig or wind speed would drive it. Other speed records could be arranged with different stipulations. The designer of *Crossbow* has been the first to admit that his concept is of scant relationship to everyday yachts and boats. Trials could be conducted over a 'there and back' run, but this would merely result in lower speed figures (*Icarus* was quite capable of going on either tack). There is the possibility of a trial to windward, which would produce boats designed to be fast when close to the wind and able to tack quickly. A difficulty here would be to say when the wind (which is frequently varying slightly in direction) had not shifted to give an advantage to a boat for a few moments, as she beat up the windward course. Conditions would seldom be right.

One of the original craft to attempt sailing speed records, *Clifton Flasher*, with multi-aerofoil sails.

Ice boats

The fastest of all speeds using sails are not on water at all, but on ice and on land. This demonstrates that it is not wind power that is the culprit in the slow absolute speed of boats, but the poor stability and 'stickiness' of floating in water. Sailing on land and ice gives high speeds because, firstly, lateral resistance is infinite, therefore the power is nearly all transmitted into forward motion; secondly, heeling resistance is enormous. Further, on ice, resistance to forward motion is very small.

Fastest speeds under sail are achieved on ice. This DN class boat has skate for forward steering and wing mast with battens.

The fastest speed sailing on ice was recorded as 143 miles per hour by a Class A stern-steerer on Lake Winnebags, Wisconsin, USA in 1938. The driver was John D. Buckstaff.

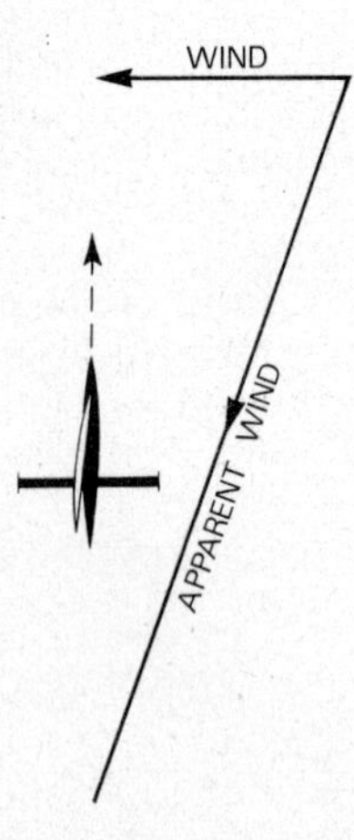

Why an ice yacht is invariably close-hauled. Her own speed causes a moderate beam wind to become a strong wind at a fine angle.

The largest ice yacht to be built was for John E. Roosevelt of the New York Yacht Club in 1870. Its members used to experiment in very cold winters with some of the sails of their yachts rigged on huge ice boats. Roosevelt's 'boat', *Icicle*, was 68 ft 11 in long and had a sail area of 1070 ft². She was sloop rigged as were most ice boats of the time. Two cat-rigged boats were reported – *Advance* and *Snow Bird*. Names of sloops included *Arctic*, *Hail*, *Avalanche*, *Jack Frost*, *Whiz* and *Snow Flake*. All these ice boats sailed on the Hudson River, when frozen, under the burgee of the Poughkeepsie Ice Yacht Club.

The largest ice yacht, *Icicle*, she sailed on the Hudson River in 1870. The large size was counter productive for speed owing to weight.

The most favoured classes in modern ice yachting include the 15 metre² (refers to sail area) with two men who sit as in an old aeroplane with the steering skate aft and cross-beam forward. It is used in Europe particularly in Holland when the waterways freeze. The 15 metre² can achieve 60 mph in a 24 mph breeze.

A cheaper class is the DN which is sailed single-handed. It originated from a design competition run in 1937 by the *Detroit News*. It is 12 ft in length with sail area 70 ft² and weighs 154 lb.

Land boats

Land or sand yachting is more common, not being dependent on a hard winter. The DN is also used with wheels, though a more modern class, the Belgian DAD was introduced in 1970. It has a sail area of 78·5 ft².

The only authentic timed runs in Britain are made annually at the meeting of the Anglian Land Yacht Club. Flying kilometres and 12 hour endurance runs take place. At the 1974 meeting speeds only reached 7 mph showing that a record depends on the wind of the day.

The fastest speed in a land yacht was recorded in 1956 at Lytham St Anne's, Lancashire at 57·69 mph over a measured mile. The yacht was *Coronation Year II* owned by R. Millett Denning and sailed by J. Halliday, Bob Harding, J. Glassbrook and Cliff Martindale.

The first land vehicles to be driven by sails were probably used on the plains of ancient China. Invention of land yachts in Europe belongs to the Belgians: a design of 1600 is shown in Bruges Library. (It, therefore, pre-dates yachting on the water, see Section 2.) A seventeenth-century Dutch land ship is said to have travelled 42 miles in 2 hours. In the USA in the nineteenth century Kansas Pacific Railway repair-parties used a sailing-car on the rails which made an average of 30 mph and it was claimed that a record distance under sail of 84 miles was sailed in 4 hours.

DN and A class land yachts.

Modern land yacht with fully battened single sail and wide wheel base.

The first development of a land yacht as a sport was by Frank and Ben Dumont in 1910 at Bruges, Belgium. Using bicycle wheels and running on the hard sand beaches of the area, the sand yacht travelled faster than motor vehicles of the time. In 1925 spoked wheels were introduced and the basic three-wheel design (with single wheel either forward or aft) became as it is now.

Donald Campbell

UNDER POWER

The highest speed ever observed on the water was by Donald Malcolm Campbell in *Bluebird K7* on Coniston Water in the English Lake District, just before 8·50 am on 4 January 1967 when he achieved 328 mph just before the boat fatally disintegrated. *Bluebird* was a $2\frac{1}{4}$ ton turbo-jet-driven craft of 26 ft 5 in, designed by Norris Brothers with a Metro-Vickers Beryl jet, consuming 650 gallons of kerosene per hour.

The official world unlimited water speed record is held by Lee Taylor Jr of California, USA who in the hydroplane *Hustler*, 30 ft 6 in, on 30 June 1967 on Lake Guntersville, Alabama, achieved, using the propulsion of a Westinghouse J-46 turbo-jet, a speed of 285·213 mph. A braking parachute was used to slow the boat after the run.

The greatest speeds on the water utilizing various sorts of power are considerably lower than that made with wheeled vehicles and far less than

Bluebird K7, fastest ever on water.

Lee Taylor's *Hustler*, 30 ft 6 in, holder of the world unlimited water speed record.

those achieved in air and space. This will always be so because the boat is an interface vehicle (i.e. has to be propelled on the difficult boundary between air and water). This is demonstrated even more by the way in which higher speeds are actually arrived at in water transport by lifting the 'boat' off the water by planing on a very small surface of the hull or by developing a hover vehicle which when moving is not directly in contact with the water, using it as a support for a cushion of air only. The great variety in methods of propulsion means that there are now records for all sorts of differently propelled craft, classified by the means of propulsion (for example jet in air, propeller in water, propeller in air) and the capacity of the motor. These are shown below. Any comparison with speeds of boats under sail are today almost pointless. Suffice it to say that the official record under sail is about one-tenth of that under power, or some 250 miles per hour less. The fastest sail-boat is standing still when these record boats pass her. Of course, they never physically do so because the highest records are made on glassy calm water and away from other craft. Sailing speeds need wind which ruffles the water and increases the resistance. This was not always so, for the earliest speeds under power took many years before they exceeded the fastest commercial sailing ships, but then in the early part of the twentieth century they quickly entered velocities of a different order.

Patents for the first engine-propelled vessel were taken out as early as 1737. The patented vessel had rear paddles connected by belts or rope to big wheels driven by a steam-engine. The design of the boat was semi-open and the obvious purpose was to tow sailing-vessels in a calm, for instance from a dock out to a harbour mouth where the sails could find a breeze. The envisaged speed would have been about 2 mph – that of the sailing-vessel in a light air.

First power

The first successful application of steam afloat was at Lyon, France in 1783. The earliest commercial steamboats began running on the Clyde in 1801 (*Charlotte Dundas*) and Hudson River in 1807 (*Clermont*). From this time (when Nelson's captains were, therefore, aware of mechanically propelled boats usable in sheltered waters), steam was developed in commercial ships and boats. The first steam yacht to be owned by a member of the Royal Yacht Squadron was commissioned by Mr J. Assheton-Smith in 1830 and Queen Victoria's first steam-yacht *Victoria and Albert*, complete with paddle-wheels, was ready in 1843. Lack of dependence on the wind and not high speed was the objective.

To attain speed as such it was necessary to progress towards highly powered small craft. The steam engine was in practice too heavy and slow running – not to mention its coal fuel – to seek much speed. Of course, a steamer travelling across the oceans at 8 knots would usually beat a sailing ship which sailed at 15 knots in good winds, but 3 knots in a light wind. Small craft needed a different form of propulsion and the first motor boat travelled on the Seine in Paris in 1864 driven by a 2 hp gas-driven engine made by an early motor car pioneer, J. J. Etienne Lenoir.

The first motor boats in regular use were those by the German police and were built and used in Hamburg in 1890. They were wooden launches fitted with Daimler internal-combustion engines.

The first motor boats regularly built in Britain were produced in a boat-yard at Putney in 1893. Daimler engines from Germany were installed by Daimler Motor Syndicate Ltd. From then on petrol engines for small boats developed steadily.

The first sensational demonstration of speed by a powered vessel was in 1897 by the steam yacht *Turbinia*, a craft of 103 ft 4 in, displacement 44½ tons. She was designed by Sir Charles Parsons and had the first turbines in a boat. The engines drove nine propellers each of 18 in. in diameter. *Turbinia* created an international impression by steaming through the assembled international naval review for Queen Victoria's Diamond Jubilee at a speed of 35 knots (just under 40 mph).

First power races

At the turn of the century motor boat races were held off Monaco. In 1905 there were 104 entries for a meeting there.

The first successful attempt to rationalize the progress of racing and speed records was by the presentation of the 'British International Trophy' by Sir Alfred Harmsworth, later Lord Northcliffe the newspaper proprietor. It was for a race for power-boats under 40 ft. The first event was from the Royal Cork Yacht Club to Glanmuire, Ireland, an 8½ mile course, on 11 July 1903. The winner was *Napier Minor*, 40 ft, with 75 hp Napier engine, driven by Campbell Muir. The average speed was 19·5 mph and the maximum 23·5.

In 1904 the same boat was second in the first Calais to Dover race, driven by S. F. Edge. The winner was *Mercedes IV* in 1 hour 0 minutes 0 seconds – 21 boats started and 20 finished.

The first regular offshore power race in Britain was the London to Cowes event which ran from 1906 to 1938. The races were in seagoing cruiser-type vessels and various prizes were given for first to finish, own predicted times, for classes and so on. The planing offshore boats of today did not exist.

Planing boats were being discovered, however, and used in sheltered water. In 1872 the Reverend C. M. Ramus had submitted a plan to the Admiralty for a double wedge 'to cause the ship to be lifted out of the water'. In 1877 Sir John Thorneycroft took out patents for skimming vessels. In 1908 there were further patents by William Henry Fuber (an American living in France). The principles were known: the difficulty was finding a power to give the thrust to lift the boat.

International rivalry began in the Harmsworth Trophy in 1907 when the American *Dixie* owned by Commodore E. J. Schroeder, won it in Southampton Water.

The last non-planing boat to win the Harmsworth Trophy was *Dixie II* (USA) in 1908 at Huntingdon Bay, Long Island. She averaged 32 mph. She was the 40 ft maximum, with 200 hp, but weighed only 1000 lb. The course was three 10 mile triangles. This second boat of Schroeder's, designed by Clinton Crame, beat two American and two British boats. Over a measured mile she was timed at a record for a 40-footer of 36·6 mph.

Planing speeds

The rise of the planing boat, known as the *hydroplane*, came about when S. E. Saunders of Cowes, Isle of Wight, developed the patent ideas of Fuber. He put in a hull a number of longitudinal steps and gave the hull a V-section. (The term 'hydroplane' was at this time applied to all planing craft. Today planing boats are so common that the term has fallen into disuse and is reserved for boats with a longitudinal step. Multi-steps are no longer designed.)

The biggest advance in power-boat speeds before 1914 was the arrival on the scene of *Maple Leaf IV*. She was designed by S. E. Saunders as a multi-step hydroplane; there were five steps in the 40 ft hull. Beam was 8 ft 5 in, draught 1 ft 4 in. She was built of mahogany and rock elm timber.

The first speed on the water over 50 mph was achieved by *Maple Leaf IV*. She went to America in 1912 and driven by T. O. M. Sopwith regained the Harmsworth Trophy for Britain. In 1913, Sopwith defended the trophy in the same boat in the Solent at an average speed of 57·45 mph. *Maple Leaf IV* was the fastest boat in the world up to 1914. The First World War increased knowledge of high-speed craft as the world's navies developed fast torpedo-boats.

Miss England II (agency photo).

The early rivalry between Britain and America in the Harmsworth Trophy has never been repeated for in 1920 Gar Wood won the trophy in *Miss America I* and the trophy remained in North America. Gar Wood had a series of boats called *Miss America* (*I* to *X*). *Miss England I* and *Miss England II* made unsuccessful attempts to regain the trophy. In 1931 *Miss England II* raced against *Miss America VIII* and *Miss America IX* in Detroit. *Miss England II*, driven by Kaye Don, foundered in the wash of *Miss America IX*, driven by Gar Wood, enabling George Wood in *Miss America VIII* to collect the trophy. In the first heat *Miss England* had put up the best time at 89·9 mph and was probably the faster boat.

By this time the fastest speeds afloat could no longer be made in competition, but only in isolation, away from other craft and on smooth water. This was shown as these same boats in the Harmsworth races had lower speeds recorded than on their own. But other craft purely to attempt the world speed record were developed.

The progress of water speed was from 50 mph in 1914, to the first 100 mph in 1931, to the first 200 mph in 1955, to a fatal run of over 300 mph in 1967. The first jet-propelled craft to break the speed record was *Bluebird* at 202 mph in 1955, so both the 200 and 300 mph barriers were first penetrated by Donald Campbell.

PROGRESSIVE SPEED RECORD (ALL ARE BRITISH OR US)

Year	Boat	Driver	Location	Speed (mph)
1928	*Miss America VII*	Gar Wood	Detroit	92·9
1930	*Miss England II*	Sir Henry Segrave	Lake Windermere	98·7
1931	*Miss England II*	Kaye Don	Argentina	103·5
1932	*Miss America IX*	Gar Wood	Miami	111·7
1932	*Miss England III*	Kaye Don	Loch Lomond	119·7
1932	*Miss America X*	Gar Wood	Detroit	124·9
1937	*Blue Bird*	Sir Malcolm Campbell	Lake Maggiore	129·5
1938	*Blue Bird*	Sir Malcolm Campbell	Lake Hallwil	130·9
1939	*Blue Bird*	Sir Malcolm Campbell	Coniston Water	141·7
1950	*Slo-mo-shun IV*	Stanley Sayres	Lake Washington	160·3
1952	*Slo-mo-shun IV*	Stanley Sayres	Lake Washington	178·5
1955	*Bluebird*	Donald Campbell	Ullswater	202·3
1955	*Bluebird*	Donald Campbell	Lake Mead, Nevada	216·3
1956	*Bluebird*	Donald Campbell	Coniston Water	225·6
1957	*Bluebird*	Donald Campbell	Coniston Water	239·1
1958	*Bluebird*	Donald Campbell	Coniston Water	248·6
1959	*Bluebird*	Donald Campbell	Coniston Water	260·3
1964	*Bluebird*	Donald Campbell	Lake Dumbleyung	276·3
1967	*Hustler*	Lee Taylor Jr	Lake Guntersville	285·2

After 1945 the Harmsworth Trophy was raced in North America. The Canadian boat *Miss Supertest III* won the trophy in 1959, 1960 and 1961. Britain did not race for the trophy after 1945 and neither did the USA after the three Canadian successes and the trophy was never raced for again. In North America the Gold Cup has succeeded the Harmsworth Trophy, which can now be seen as a pioneer event of motor-boat racing. The fastest average speed ever attained in the Harmsworth Trophy was in 1960 by *Miss Supertest III* at 116·5 mph. Shortly after winning the 1961 event in *Miss Supertest*, Bob Hayward was killed in another motor-boat race at Detroit.

Ultimate water speeds

In the attempts on the world water speed record, the toll of British speed kings was high. The 1930 record by Sir Henry Segrave was achieved but on his subsequent run to try and record an even higher speed, *Miss England II* struck a submerged log on Lake Windermere. The day, Friday, 13 June, was a fine one, so the crew of three were not wearing life-jackets (nowadays it would be *de rigueur*). As the boat foundered Segrave and one of the two engineers were drowned. One engineer was saved and the boat was quite quickly salvaged and raced again later. Before the fatal collision *Miss England II*, 36 ft with two 1750 hp engines, became the first waterborne craft to be observed at a speed of over 100 mph.

In 1952, the second major accident occurred with the death of John Cobb, the motor-car speed record-holder, in an attempt on the world speed record of *Slo-mo-shun*. The boat used was the single-step *Crusader* designed and built by Vosper of Portsmouth. John Cobb began his final run at 11.55 on 29 September. *Crusader* reached an observed speed of 206 mph, far in excess of any contemporary rival, but then performed one or two slight 'porpoise' variations in course. Going at maximum revolutions, the craft quite suddenly dived and disappeared in a cloud of steam. A few moments later only wreckage was visible and the body of Cobb (speed records were by now solo affairs) was picked up from the water having surfaced due to a life-jacket.

Bluebird K7, driven by Donald Campbell, achieved the 328 mph, the highest ever speed on water after about a month of trials on Coniston Water in the English Lake District. Travelling from south to north on 4 January 1967 the boat's starboard sponson left the water, hit it again, flew off it and then the entire boat somersaulted, coming down in the water in a cloud of spray. The boat's nose piece appeared to hit the water and the craft disintegrated, both sponsons flying apart from the hull structure. The main hull sank, as did the body of Campbell which was never recovered. This was the last British attempt on the unlimited world water speed record.

Records

The most complex classification of speed records has come about in the last decade with trials and races graded by type and engine power. Speed records, hour records, distance records and competition records are all laid down from time to time by the Union International Motonautique (UIM), or the American Power Boat Association. To increase a record, a margin of at least 1·0075 second over the existing record is needed. Timing is up to one-tenth of a second up to 200 kilometres per hour and up to one-hundredth for higher speeds. Records of the UIM are always stated in mph (statute miles) and k/hour (kilometres) though the distances used may be nautical miles.

As well as the different types of record, the boats are split up into racing inboards, sport inboards (which may have diesel engines, turbo-jets, rocket propulsion, aerial propellers or water propellers), racing outboards, inflatables, and offshore boats of various types.

Inshore racing outboard by Jeff Edwards.

Records in many classes are continually creeping up. Those below are given to show the sort of speeds attained and are up to 21 October 1974. Years are given and these show a period without any progress in some cases. The inshore boats have a number of classifications for various capacities of power (for example 850 to 1000 cc, then 1000 to 1500 cc and so on) but those given here are for up to 1000 cc and also the group with the best speed of the type. Boats which elect to try for the speed record of a group cannot hold any other speed record by claiming the speed for a group of greater power which has remained slower. Boats with turbo-jets, rocket propulsion and aerial propellers can only engage in solo speed trials and cannot for obvious reasons engage in close competition. All records given are world; national ones are recorded by individual national authorities.

Offshore classes	**mph**	
Class 1 (up to 16400 cc petrol)	85·63	Countess of Arran (1971)
Class 2 (up to 8200 cc petrol)	81·79	Countess of Arran
Diesel propulsion	78·36	M. Livio (1924)
Inshore classes		
Racing inboards:		
up to 1000 cc	94·98	C. Casalini (1971)
2000 to 2500 cc	131·12	Foresti Franco (1970)
Sport inboards:		
up to 1000 cc	82·16	K. Bonikowski (1973)
1500 to 2000 cc	104·78	Gilberti Franco (1971)
Racing outboards:		
up to 1000 cc	110·48	H. Entrop (1969)
1500 to 2000 cc	136·38	James F. Mezten (1973)
Sports outboards:		
850 to 1000 cc	75·54	Joe Marteens (1972)
1000 to 1500 cc	80·31	F. Miles (1970)
Inflatable boats:		
500 to 700 cc		
min. weight 60 kg; 3·7 metres length	56·56	T. Williams (Britain) (1973)
Aerial propeller, weight 1200 kg	96·86	Venturi (1951)
Limited jet propulsion, weight 1200 kg	162·63	T. Watts (1969)
Unlimited class, but immersed propeller	202·42	Larry Hill (1973) in *Mr. Ed* a supercharged hydroplane off Long Beach, California

Unowot, winner of the Cowes–Torquay–Cowes race 1973.

ABSOLUTE SPEED: HOW FAST OR SLOW BOATS REALLY GO

Speed (knots)	Circumstance
2·9	Boat speed of 7 ft 8 in Optimist sailing-pram in favourable conditions.
5	Minimum speed demanded for sailing yacht of 25 ft LWL under power for engine allowance in rating rule.
5·7	Typical average of 35 ft ocean racer on week-end offshore event with mixture of weather, 1975.
6·5	Average speed on voyage of a nineteenth-century clipper ship.
7·82	*Great Britain II*, 77 ft ketch, average on Round the World race, 1974.
8·01	*Ranger*, J class sloop, dead to windward on 10 mile leg, 1937.
8·05	Fastnet race record by *American Eagle*, 68 ft sloop, in 1971, average over 605 mile course.
8·33	*Manureva*, 70 ft trimaran, when single-handed on 400 mile run in Indian Ocean, 1973.
8·5	Round Britain race record by *British Oxygen*, 70 ft catamaran.
10	Probable absolute maximum of any vessel, sail or oars, before 1800.
10·7	*Superdocious*, Gold Medal winning, Flying Dutchman planing dinghy over recorded 500 m course.
10·72	*Windward Passage*, 73 ft ketch on Miami–Jamaica race, average over 811 mile course.
12·5	*Britannia*, LWL 86 ft 9 in, cutter in English Channel during race in 1896.
13·58	*Manureva*, best 24 hour run, 1973.
13·75	*Preussen*, 433 ft, five-masted ship with 8000 tons of cargo in 38 knot beam wind, 1902.
17	40 ft speed-boat over Harmsworth race coastal course in 1903.
17	*Satanita*, cutter (LWL 93 ft 6 in) in smooth water broad reach on the Clyde, 1893.
19·5	*Champion of the Seas*, clipper ship, alleged average speed in a 24 hour run.
19·6	Tornado class catamaran in class trim, over recorded 500 m course, 1972.
31·09	*Crossbow*, 55 ft proa. Fastest sailing speed ever on water authentically recorded, 500 m course, 1975.
35	*Turbinia*'s demonstration speed in 1897.
50·7	Fastest land yacht, *Coronation Year II*, 1956.
59	Record over Cowes–Torquay–Cowes race course, 1974.
74	Record for Class I offshore power-boat speed.
125·84	Fastest ice yacht, Class A stern-steerer, 1938.
250	World water speed record – *Hustler*, 1967.
288	Highest speed ever observed to be achieved on water – *Bluebird*, 1967.

Inshore powerboats, Formula I at the Bristol Grand Prix.

BOVRIL
402
Mobil
9
KIEKHAEFER
aeromarine·IX
MARTINI
RACING
9
Dry Martini

Top left
Hot Bovril, winner of the Cowes–Torquay–Cowes race 1971.

Middle left
Aeromarine IX, winner of the Cowes–Torquay–Cowes race 1972.

Bottom left
Fastest speed over 236 miles on the Cowes–Torquay–Cowes race in 1974 was by *Dry Martini* driven by Carlo Bonomi.

Right
Winner of the only Round Britain powerboat race in recent years, *Avenger Too*, victor of the *Daily Telegraph* race of 1969.

The longest race for power-boats off Britain is the Cowes–Torquay–Cowes race of 236 miles, sponsored by the *Daily Express*. The race is held annually. The course originally finished at Torquay.

The first race on the course was in 1961, when there were 27 starters between 18 and 40 ft WL. On a 160 mile course which had 'dog legs' in it to bring it near spectator points on shore, the winning boat averaged 22 knots. This was *Thunderbolt* owned by T. E. B. 'Tommy' Sopwith (son of Sir Thomas Sopwith, see Section 3), 30 ft, designed by Raymond Hunt (USA) and with a 650 hp engine. Sopwith won the race again in 1968 and 1970 and is the only person to have won it three times.

The fastest average speed for the 236 mile course has grown almost yearly since 1961. The record is by the winner of the 1975 race when *Uno-Embassy* (Hyams/Hoare/Shead), 37 ft, a Class I offshore power-boat with 1200 hp engine covered the 236 mile course at an average speed of 63·24 knots, finishing 40 minutes before the next boat. The previous best was *Dry Martini* (USA) in 1974 at 58·9 knots.

The longest power-boat race in Europe was the 2947 mile London to Monte Carlo in June 1972. It was taken with 13 stopping-points *en route*. The winner, *HTS* (Britain), averaged 36·3 knots. In 1969 the *Daily Telegraph* sponsored a power race round England from Portsmouth and back via the Caledonian Canal in ten stages. The winner from 42 starters was A. Pascoe Watson's *Avenger Too* – 24 boats finished the course.

The only world drivers' championship in offshore power-boat racing is for Class I boats. Under the auspices of the UIM, the award is the Sam Griffith Memorial Trophy named after the man who was prominent in getting offshore power-boat racing started after the Second World War. Points are awarded for the first six finishers in a number of nominated races and the best seven are counted: the rules in this respect may vary slightly from time to time. The champions have been as follows.

Year	Name	Nationality	Type of craft	Engines
1964	Jim Wynne	USA	Coronet	2 Volvos
1965	Dick Bertram	USA	Bertram	Detroit diesel
1966	Jim Wynne	USA	Souter	2 Daytonas
1967	Don Aronow	USA	Magnum	3 Mercury outboards
1968	Vincenzo Balestrieri	Italy	Magnum	2 Mercruisers
1969	Don Aronow	USA	Cary	2 Mercruisers
1970	Vincenzo Balestrieri	Italy	Cary	2 Mercruisers
1971	Bill Wishnick	USA	Cigarette	2 Mercruisers
1972	Bobby Rautbord	USA	Cigarette	2 Mercruisers
1973	Carlo Bonomi	Italy	Cigarette	2 Aeromarines

2 Racing under Sail

ORIGINS

The word 'yacht' originated in Holland in the seventeenth century. In Old Dutch *jagen* meant to hunt or chase and *jaght* meant hunting or chasing. So *jacht-schip* meant a fast vessel or a vessel for a sportsman. A Dutch dictionary of 1573 shows that the word did not exist even in Holland. In 1660 in England, Sir Anthony Deane said to Samuel Pepys, 'we had not heard of such a name as "yacht" in England'.

The first use of yachts was in Holland to convey 'persons of rank' on the many waterways of that country. Distinctions, therefore, arose between a yacht and any other vessel. Most craft afloat at that time were uncomfortable and today would be considered a hardship in which to sail. But a yacht was smart and clean in a manner introduced by the Dutch, had fittings of comparative luxury and was reasonably comfortable afloat as demanded by the increasing prosperity of the users. After 1660 in England, yachting became recognized because, most important of all, *yachts were patronized by the King*.

Yachts were introduced to England by King Charles II after his restoration in 1660. Before that time royalty had, of course, travelled by water in vessels of various size and performance, but such ships were not permanently intended for pleasure and cannot be regarded as connected with pleasure sailing or yachting, even if the sovereign or other persons enjoyed sailing in them.

The first sail on a yacht by Charles II was on the first part of his journey back to England from Holland for the Restoration. He travelled from

The first picture of yachts racing or at least rallying. This early 17th-century painting (in the National Maritime Museum, Greenwich), signed by Andries van Ertbelt, shows Dutch yachts racing before the word 'yacht' was known outside Holland. It looks here as though a sudden squall has struck the fleet. The yachts of Charles II were English elaborations of these sailing boats.

The present Starcross Yacht Club on the Exe Estuary. Its first regatta was in 1772.

Breda to Delft and so enjoyed it he determined later to have a boat of his own. This is the first recorded example of someone catching the 'sailing bug'.

The first yacht introduced into England was a gift from the Dutch to Charles II and built in Amsterdam. Called *Mary*, she was 52 ft in length, beam 19 ft, draught 10 ft. The crew was 20 men in peacetime and she carried 8 guns. She arrived in London River in August 1660. On 15 August, Samuel Pepys reported in his diary 'To White Hall, where I found the King gone this morning by 5 o'clock to see a Dutch pleasure boat below bridge.'

The first yachts built in England soon followed as English builders imitated and attempted to outclass the Dutch boat. In November Pepys went on board *Mary* with the established builder Mr Pett. 'In the afternoon, I went on board the yacht, which indeed is one of the finest things that ever I saw for neatness and room in so small a vessel. Mr Pett is to make one to outdo this for the honour of his country, which I fear he will scarce better.' Ever since then designers and builders have continued to claim that they can produce improved and newer boats. So Mr Pett built *Jenny* for the King and *Anne* for his brother the Duke of York.

The first yacht race took place on 1 October 1661 between the yachts of the King and the Duke of York. It was from Greenwich to Gravesend and back on the River Thames. It was for a wager of £100. The King lost going, but won on the return. It was sailed in an east wind and the Duke of York's vessel must have been better to windward. On the King's yacht 'there were diverse noble persons and lords, his Majesty sometimes steering himself. His barge and kitchen-boat attended.' The *Mary* was eventually wrecked near Holyhead in 1675 and was located under water in 1973.

Oldest clubs

The first known regatta conducted by a club was in 1720 when The Water Club of the Harbour of Cork (south-west Ireland), issued rules for its members: it was 'ordered that The Water Club be held once every spring tide in April to the last in September, inclusive'. A painting by Monamy

Hornet class. Largest British national class to be used behind the Iron Curtain

American Eagle, fastest boat round the Fastnet course

Optimist class. These Japanese children swell the thousands that sail this tiny boat

in 1738 shows the club at manœuvres, a few leagues out at sea, led by an Admiral. There was no racing, but what nowadays would be called a rally. 'The fleet fall in their proper stations and keep in their line in the same manner as the King's ships. This fleet is attended with a prodigious number of boats, which, with their colours flying, drums beating and trumpets sounding form a most agreeable and pleasant sight.' After 1765, the activities of the club lapsed, or are at least unrecorded. In 1828 the Cork Yacht Club was formed and today exists as the Royal Cork Yacht Club. This claims ancestry of The Water Club and, therefore, its foundation as 1720, and to be the world's oldest yacht club.

The oldest yacht club in Britain which is still in existence is the Starcross Yacht Club at Powderham Point on the Exe Estuary. The club now has 550 members and races centreboard dinghies. The first regatta of the club was held in 1772 from Powderham Point where a quay and slipway were in use for barges bringing in supplies locally. The leaders of the regatta were Lord Courtenay and Sir Lawrence Palk of Haldon: as with all these early regattas the organization was in the nature of an annual event including an 'Admiral' leading a fleet of boats. A contemporary newspaper mentions 'the Starcross Club' in connection with the 'Fete Marine' of 14 August 1775. Early records of the Starcross Yacht Club were lost and such items as cups presented 'by the Stewards of Starcross Yacht Club' are the main clues to what may have been an intermittent life. Certainly between 1922 and 1933 the club was amalgamated with another, but in 1933 it opened up again at the village of Starcross. In 1957 the club transferred back to its place of origin at Powderham Point a few miles upstream, as this was more suitable for hauling up modern centreboarders.

The first regatta to be called such and held in England was at Ranelagh on the Thames on 23 June 1775. 'Regatta' was an Italian word and water festivals had been derived from Venice with its waterways. Proceedings were confined in this first regatta to 'watermen', but 'sailing boats and pleasure vessels belonging to several very respected gentlemen were present'.

The first English yacht race for other than royalty was on 13 July 1775. It was for a Silver Cup presented by the Duke of Cumberland for a race for sailing-boats from 2 to 5 tons 'never let out for hire', from Westminster Bridge to Putney Bridge and back. The cup was won by *Aurora* owned by Mr Parks of Ludgate Hill.

The first English yacht club to be subsequently a 'royal' club was named the 'Cumberland Fleet', which was formed as a result of this Thames yacht race and subsequent races on the Thames. Each year the Duke of Cumberland gave a cup and races were generally between Putney and Blackfriars. The Cumberland Fleet continued until 1823, when on the coronation of King George IV, it became known as the 'Coronation Society'. As the result of a dispute over a protest a splinter group formed The Thames Yacht Club: this was the first occasion to demonstrate the seriousness with which yacht racing can be taken. The Coronation Society lapsed and the new club became the Royal Thames Yacht Club in 1830. The club has a continuous history and its activities moved away from Thames waters with increasing industrialization.

The club has remained very active into the 1970s and has still a London club house at 60 Knightsbridge and also a sailing station and club house on the Hamble River.

The Royal Thames Yacht Club today is at Knightsbridge, London.

Right
The Royal Yacht Squadron today.

The first sailing match outside the Thames is recorded by a print of a regatta at Cowes in 1776. Only watermen took part, probably fishermen and pilots having 'working boat' races on holidays and regatta days. In 1788 there was a 'sailing match for thirty guineas to take place at Cowes for vessels carvel built not exceeding 35 tons register, westward round the Island'.

The earliest recorded foul in a sailing race was in *The Times* of 23 July 1795. It was between two boats, *Mercury* and *Vixen*. The captain of the *Vixen* cut away the rigging of the rival yacht with a cutlass and 'dismantled her'.

The enthusiasm of gentlemen yacht-owners was aroused by the local races and regattas and the scene was set for the beginning of yachting organized for and by them for its own sake. It is important to appreciate that the recurrent wars of the eighteenth century, mainly against France, made the seas unfit for sailing except in strict necessity. Even Turkish pirates would lie at times a few miles off the coast of England. The yacht *Hawke* of the Cumberland Fleet which had ventured into the Straits of Dover in 1777 was chased into Calais by an American privateer. The end of the wars and a period of prolonged peace began in 1815 with the downfall of Napoleon at the Battle of Waterloo. Freedom of the seas was ensured by the British Navy.

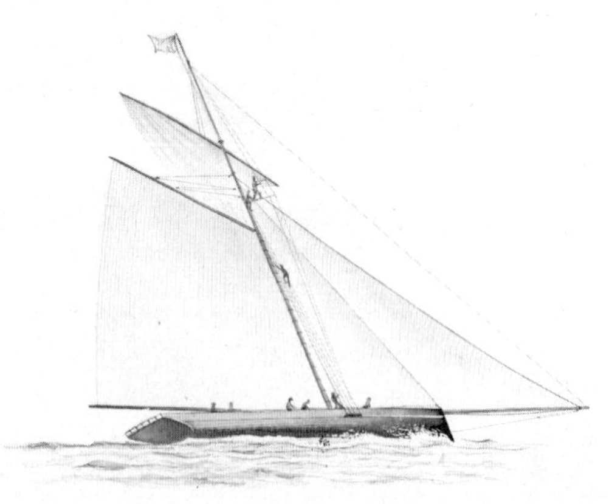

The 85 ton cutter *Arrow* handing her topsail. Owned by Joseph Weld from 1821 to 1828, she was the first winner of a race organized by the Royal Yacht Club.

Typical of early Victorian clubs, the Royal Southern Yacht Club still stands in Southampton, where it was established in 1838. But the building is commercial, the club members having fled many years ago to a new site on the Hamble River.

Right
The Southern Yacht Club at New Orleans. Here is part of the 125-year anniversary celebrations.

The year 1815 marked the formation of the first yacht club organized in the sense that we know today. On 1 June, 42 members of the aristocracy and gentlemen met at the Thatched House Tavern, St James's Street, London and formed what was called 'The Yacht Club'. From then on several races a year were held at Cowes, in the Solent and sometimes round the Isle of Wight. At first these were matches between private owners, but in 1826 the club introduced races for cups and prizes during its annual two-day regatta on Thursday, 10 August and Friday, 11 August. Thus the progression of organized yacht racing began. The winner of the first race to be held in 1826 was *Arrow* owned by Joseph Weld; she was thus the first winner of a race organized by a royal club for a cup (as opposed to a match or wager). In 1820 the club became The Royal Yacht Club and on 4 July 1833 it was named the Royal Yacht Squadron with the approval of King William IV. In 1858 the RYS occupied the Castle at West Cowes, which has been its base ever since.

John Cox Stevens

Origins in the USA

The first yacht race in the United States of America was more informal. The schooner *Sylph* owned and sailed by John P. Cushing of Boston rounded Cape Cod on 3 August 1835 and encountered *Wave* owned by John C. Stevens of New York: a race followed and then two further races were arranged in Vineyard Sound.

The 'father of American yachting' is John Cox Stevens, member of an inventive, enterprising and subsequently wealthy New Jersey family. In his yacht *Gimcrack* anchored off the Battery on the south of Manhattan, meeting at 5 pm on 30 July 1844, a group of men formed the New York Yacht Club. Its first regatta was on 16 July 1845 and nine yachts (six schooners and three sloops) of between 17 and 45 tons took part. The first yacht to finish was *Cygnet* owned by William Edgar. Annual regattas took place thereafter, so when an invitation came from the British to send a yacht to race on the occasion of the year of the Great Exhibition in 1851, the New York Yacht Club was well able to handle the challenge.

The oldest club in the United States to remain in active waterfront premises (the New York Yacht Club is at 37 West 44th Street in Manhattan) is the Southern Yacht Club, which is also the second oldest club in America, being founded in 1849. The club is at New Orleans, Louisiana. The first club house was on the coast of the Gulf of Mexico, but it soon moved to the shores of the large Lake Ponchartrain. The Victorian club house built there was in 1950 replaced by a modern building.

The New York Yacht Club now at West 44th Street, Manhattan. It moved into this Victorian building at the turn of the century and remains there among the skyscrapers.

France

The first yacht club in France was the Société des Régates du Havre: this club is still active today. It is well frequented by British yachts which cross the Channel and the Secretary sometimes tells them that the Société is a 'royal club' having been formed under the patronage of King Louis-Philippe in 1840. The formation followed the first yacht race in France when 21 vessels raced off Le Havre in a regatta in 1839. One other yacht club had been formed in continental Europe before this: the Kungl. Svenska Segel Sallskapet – the Royal Swedish Yacht Club – in 1830.

In Paris, the Cercle de la Voile de Paris was formed on 7 March 1858 (originally under the name of Cercle des Voiliers de la Basse-Seine). Regattas were held at Argenteuil on the Seine. One of the greatest founders of French yachting was Lucien More who was President of Régates Parisienne in 1859 and first Secretary of the Yacht Club de France in 1860. By 1881 the CVP had 117 yachts and 200 active members. However compared with Britain and America, France and other continental countries had few yachts and yachtsmen and the sport was hardly known except to enthusiasts.

By 1844 there were 12 yacht clubs in Great Britain and Ireland. All these had gentlemen as members, most had substantial club houses and the members owned seagoing yachts. (This 12 does not take into account numerous local boating clubs and regatta committees, who might offer sailing and rowing races.) Eleven of these were royal clubs. Apart from the Squadron, there were the Thames, Western (Plymouth, 1827), Cork, Northern (1824), Southern (1838), Harwich, Eastern (Edinburgh), Victoria (Ryde), Kingstown (Dublin) and Mersey. There was also the Arundel Yacht Club (on the Thames). Eight of these clubs are still active in the 1970s under the same names. At this time, however, the bigger yachts were run like small warships with professional crews; indeed the crews and even yachts were in practice a naval reserve, for it was still a sailing Navy then. For instance in 1844 members of the Royal Thames Yacht Club owned 146 yachts with a total tonnage of 4400, employing 540 men. The yachts ranged from schooners of 217 tons to a number of cutters between 5 and 10 tons which are the size of today's family sailing-yacht. All these clubs had one or more regattas a year. Each club had its own rules of racing, time allowances and procedure. The sport was, therefore, established on regular lines in each club by the time the yacht *America* arrived at Cowes, looking for competition in 1851.

MAJOR EVENTS

The America's Cup is long established as the most important of all yacht-racing events. It is the oldest international trophy of any sport, having been sailed for first in 1851. It has the longest unbroken sequence of wins by a single nation or club (the New York Yacht Club). It is recounted in Section 3.

The Olympic Games have included yacht racing since 1908. The classes are changed from time to time. Details start on p. 98.

The 49 ft *Aura*, world ocean-racing 'champion'.

Ocean racing

In ocean racing each trophy, series or event stands on its own merits. There is no single recognized world championship, owing to the time and cost required to move a yacht from race to race round the world. The St Petersburg Yacht Club, Florida, was the first organization to run an unofficial world championship of ocean racing. Beginning in 1969, 19 races were nominated over a three-year period. It was necessary to sail in 7 of these, including the St Petersburg–Fort Lauderdale and Miami–Jamaica events. Others included the Bermuda race, Fastnet race and Middle Sea (Mediterranean) race. Winner of the first championship was the converted (to an ocean racer) 12-metre *American Eagle*, owned by Ted Turner (USA). The second three-year series ending in 1974 and sponsored by *Yachting* magazine went to Wally Stenhouse (USA) of Chicago, sailing the 49 ft aluminium sloop *Aura*, built in 1971 and designed by Sparkman and Stephens.

The Ton Cups

Among ocean-racing events, the most significant is the One Ton Cup. This is because the yachts race level (i.e. without time allowance); it features long races and Olympic-type courses (see p. 102) and competing yachts represent the most modern development under the International Offshore Rule.

The cup itself is the most impressive of all yachting trophies being 2 ft 9 in high and 2 ft 2 in between handles and worth £4000. It was chiselled from 10 kg of solid silver by the Paris goldsmith Robert Linzeler in 1898 and paid for from the proceeds of the sale to Baron Edmund de Rothschild of the yacht *Esterel*, which had previously raced in Cannes for the Coupe de France but been defeated by the British *Gloria*. The cup was presented to the Cercle de la Voile de Paris (CVP) by the *Esterel* syndicate rather than accept the proceeds from the sale of the yacht.

The first race was in 1899 at Meulan, Paris on the Seine between the successful French defender *Belouga* and the British challenger *Vectis* of the

Island Sailing Club, Cowes. Both were small day keel boats to the French tonnage rule of 1892, and rating at or less than One Ton (see p. 43). This is the origin of the name 'One Ton Cup', the original name of which was La Coupe de CVP.

From 1899 to 1903 the race was sailed each year between France and England and once additionally Italy. Twice it was sailed at Cowes. It was sailed in One Tonners again in 1906. From 1907 it was sailed by yachts of several nations and to the then new International 6-metre rule. The One Ton Cup was raced for in 6-metres every year except 1914–19, 1940–5, 1963 and 1964.

For 1965 new rules were adopted for a combined inshore and offshore series for yachts of 22 ft rating under the RORC rule (Royal Ocean Racing Club – see p. 48). Eight nations took part at Le Havre, France. During 1970–1 the International Offshore Rule superseded the RORC rule and eligibility was changed to yachts of 27·5 ft. rating IOR.

Belouga, the first yacht ever to defend the One Ton Cup against international competition. She rated one ton under the French tonnage rule of 1892 and the race was a match against the British challenger on the River Seine.

The One Ton Cup, most elegant and also probably most competitive of all yacht-racing trophies.

WINNERS SINCE 1965

Date	Venue	Yacht	Nationality
1965	Le Havre	*Diana*	Denmark
1966	Copenhagen	*Tina*	USA
1967	Le Havre	*Optimist*	Germany
1968	Heligoland	*Optimist*	Germany
1969	Heligoland	*Rainbow*	New Zealand
1971	Auckland	*Stormy Petrel*	Australia
1972	Sydney	*Wai-Aniwa*	New Zealand
1973	Porto Cervo	*Ydra*	Italy
1974	Torquay	*Gumboots*	Britain
1975	Newport	*Pied Piper*	USA

The greatest number of competitors to take part before 1965 was 9 (6-metres) at Sandhamn, Sweden in 1930.

The greatest number of competitors to take part (since 1965) was 34 at Torquay, England in 1974.

The races now consist of three Olympic or inshore courses, one race 150 miles minimum, one race 250 miles minimum.

Far right
Close start of a One Ton Cup race, Torquay, 1974.

Gumboots, One Ton Cup Champion, 1974, and first British winner since the cup was used for ocean-racing yachts.

Two Ton Cup winners

1967 *Airela* 1968 *La Meloria* 1971 *Villanella* 1972 *Locura*
(all these were Italian and all sailed in Italy)
1974 *Aggressive* (USA) 1975 *Ricochet* (USA)

Three-quarter Ton Cup winner

1974 (at Miami) *Swampfire* (USA)
1975 (at Hanko) *Solent Saracen* (Britain)

Half Ton Cup winners

1966	La Rochelle	*Raki*	France
1967	La Rochelle	*Safari*	France
1968	La Rochelle	*Dame d'Iroise*	France
1969	Sandhamn	*Scampi*	Sweden
1970	Sandhamn	*Scampi*	Sweden
1971	Portsmouth	*Scampi III*	Sweden
1972	Marstrand	*Bes*	Denmark
1973	Hundested	*Impensable*	France
1974	La Rochelle	*North Star*	Germany

Quarter Ton Cup winners

1967	Breskens	*Defender*	Belgium
1968	Breskens	*Pirhana*	Holland
1969	Breskens	*Listang*	Germany
1970	Travemunde	*Fleur d'Ecume*	France
1971	La Rochelle	*Ecume de Mer*	France
1972	La Rochelle	*Petite Fleur*	France
1973	Weymouth	*Eygthene*	USA
1974	Malmo	*Accent*	Sweden
1975	Le Havre	*45 South*	New Zealand

Half Ton Cup yachts battle it out in the Baltic.

Right
The quarter-ton Tequila design, popular for quarter ton racing. These boats are the smallest to stay at sea overnight: note the size of the crew in comparison to the boat and its gear.

A German Scampi class half-tonner. Yachts of this design (by Peter Norlin of Sweden) scored a record three successive wins in the Half Ton Cup.

The Two Ton Cup, Three-quarter Ton Cup, Half Ton Cup and Quarter Ton Cup are derivatives of the One Ton Cup. The terms 'Half', 'Quarter' and so on do not relate numerically to any dimension of the yachts. They merely imply bigger or smaller, the length of courses being suitably adjusted. Each trophy is allotted to a size of yacht classified under the IOR as follows: Two Ton is 32 ft (rating); Three-quarter Ton is 24·5 ft; Half Ton is 21·7 ft and Quarter Ton is 18 ft. The Half Ton Cup was first raced in 1966. The Two Ton and Quarter Ton were first raced in 1967. The Three-quarter Ton was first raced in 1974.

The greatest sequence of wins by any owner in any of the Ton Cups has been three times by Peter Norlin sailing *Scampi* and *Scampi III* in the Half Ton Cup, 1969–71. In 1974 Norlin sailing his design *Accent* won the Quarter Ton Cup.

Aggressive, designed and sailed by Dick Carter, American winner of the 1974 Two Ton Cup.

Olin Stephens: the world's foremost yacht designer. He has more Fastnet and SORC winners than any other. His designs have won scores of other events since 1931, while they are as formidable as ever in 1975.

British waters

The Fastnet race is the oldest regular ocean-racing event in British or European waters. It was first raced in 1925 starting at Ryde, Isle of Wight, on 15 August. The finish was at Plymouth and after the race the competitors formed the Ocean Racing Club, now the Royal Ocean Racing Club, London. The race was sailed in 1925, then each year until 1931, then biennially from 1933 to 1939, and starting again in 1947. The race is open to seagoing yachts classified under a current rating rule and time allowance; these and the size eligibility have been altered from time to time. The course is now from Cowes to the Fastnet Rock of south-west Ireland (rounding it on either hand), then to the Scilly Isles, leaving them to port and thence to finish at the western end of Plymouth breakwater. This is a distance of 605 miles. The only yacht to win the race three times is *Jolie Brise* in 1925, 1929 and 1930. The most successful designers of Fastnet winners are Sparkman and Stephens of New York, who have designed

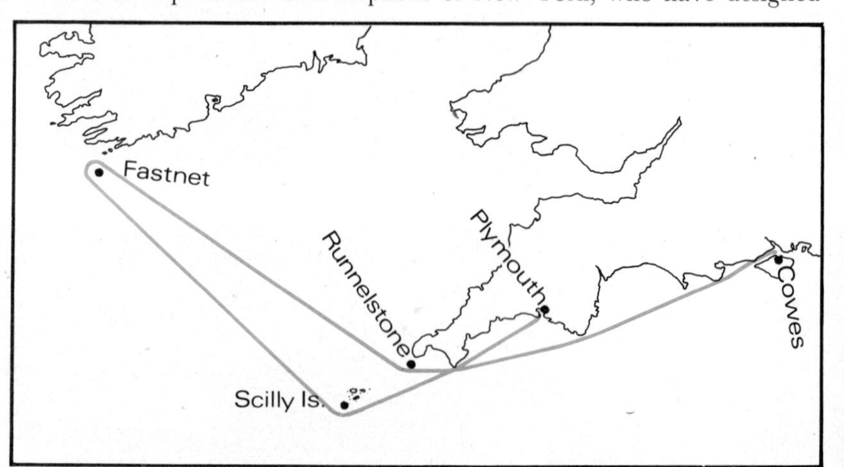

The biennial Fastnet course, which takes the yachts from Cowes round Land's End and the Fastnet Rock and back outside the Scilly Isles to Plymouth: 605 miles.

Far left to far right
Jolie Brise, first winner of the Fastnet race. In those days there were few rules about spinnakers.

Dorade, first Fastnet winner designed by Sparkman and Stephens.

Favona, designed by Robert Clark, won the Fastnet in 1953. She was the last British-designed winner and smallest yacht ever to win the Fastnet over all.

American Eagle. A 12-metre converted to ocean racing, she holds the Fastnet course record.

The Fastnet race finishes at Plymouth breakwater.

FASTNET WINNERS (CORRECTED TIME)

Year	Yacht	Designer's nationality	Owner	Nationality
1925	*Jolie Brise*	France	Lt Cdr E. G. Martin	Britain
1926	*Ilex*	Britain	Royal Engineer Yacht Club	Britain
1927	*Tally Ho*	Britain	Lord Stalbridge	Britain
1928	*Nina*	USA	Paul Hammond	USA
1929	*Jolie Brise*	France	Robert Somerset	Britain
1930	*Jolie Brise*	France	Robert Somerset	Britain
1931	*Dorade*	USA*	R. Stephens	USA
1933	*Dorade*	USA*	R. and O. J. Stephens	USA
1935	*Stormy Weather*	USA*	P. Le Boutillier	USA
1937	*Zeearend*	USA*	C. Bruynzeel	Holland
1939	*Bloodhound*	Britain	Isaac Bell	Britain
1947	*Myth of Malham*	Britain	Captain J. H. Illingworth, RN	Britain
1949	*Myth of Malham*	Britain	Captain J. H. Illingworth, RN	Britain
1951	*Yeoman*	Britain	O. A. Aisher	Britain
1953	*Favona*	Britain	Sir Michael Newton	Britain
1955	*Carina*	USA	Richard S. Nye	USA
1957	*Carina*	USA	Richard S. Nye	USA
1959	*Anitra*	USA*	S. Hansen	Sweden
1961	*Zwerver*	USA*	W. N. H. van der Vorm (O. J. van der Vorm)	Netherlands
1963	*Clarion of Wight*	USA*	D. Boyer and D. Miller	Britain
1965	*Rabbit*	USA	R. E. Carter	USA
1967	*Pen Duick III*	France	Eric Tabarly	France
1969	*Red Rooster*	USA	R. E. Carter	USA
1971	*Ragamuffin*	USA*	S. Fischer	Australia
1973	*Saga*	USA*	E. Lorentzen	Brazil
1975	*Golden Delicious*	New Zealand	P. Nicholson	Britain

* Designed by Sparkman and Stephens.

1973 Fastnet winner, *Saga* (Brazil), also designed by Stephens.

9 out of 25 winners to 1973. Their first winner was *Dorade* (USA) in 1931 and the most recent *Saga* (Brazil) in 1973. The last winner to a British design was *Favona* (designed by Robert Clark) in 1953.

The first race in 1925 had 7 starters: the largest number of starters was 258 in 1973. The fastest time for the course was in 1971 by the 68 ft ex-12-metre *American Eagle* (USA, Ted Turner) taking 3 days 7 hours 11 minutes 48 seconds (8·05 knots). The fastest time to the Fastnet Rock was in 1974 by the 61 ft 3 in *Sorcery* (USA, J. F. Baldwin) taking 48 hours 20 minutes from Cowes.

The German yacht *Nordwind* (LOA 85·2 ft) held the Fastnet course record of 3 days 20 hours 58 minutes (6·69 knots) from 1939 until beaten by *Gitana IV* (LOA 90 ft 5 in), owned by Baron Edmund de Rothschild, in 1965 when her elapsed time was 3 days 9 hours 40 minutes (7·6 knots).

The Channel race has been a regular RORC event since 1928. The course is from Southsea to the Royal Sovereign Tower (previously the lightship) to Le Havre lightship (or buoy) to Southsea. Distance is 225 miles. The first two races (1928 and 1929) used Cherbourg breakwater instead of Le Havre lightship. Eligibility rules are similar to the Fastnet and they apply in general to all RORC events. The fastest time for the course is 25 hours 49 minutes 13 seconds in 1973 by *Sorcery* (8·7 knots). The largest number of yachts to take part was 187 in 1973.

The Cowes to Dinard race (RORC) has been sailed annually since 1930 (except 1940–4). The course is from Cowes, eastward out of Spithead, then leaving all the Channel Islands to port (and in recent years the Shambles lightship near Portland Bill), thence to the entrance of the River Rance on which are Dinard and Saint-Malo. The race is for the King Edward VII Challenge Cup which the King offered in 1906 to the Club Nautique de la Rance for a race from Cowes to Dinard for 'yachts of over 30 tons'. A race between large yachts was sailed in 1913 and won by a German yachtsman who returned the cup to the French club on the outbreak of war. The cup was then unused until a suitable type of yacht and club (i.e. the ocean-racing yacht and the RORC) was in use. The largest number of starters in the Dinard race was 214 in 1973. The race was the first RORC event to take place after the Second World War, starting at Cowes on 13 September 1945, 8 yachts competed and were obliged to navigate along swept channels of minefields. The race is now sailed in conjunction with the Yacht Club de Dinard.

The Somerset Memorial Trophy was presented to the RORC in 1965 by Captain Paul Hammond USNR (ret.) in memory of Robert Somerset, DSO, one of the founder-members of the RORC and one-time owner of *Jolie Brise*. The trophy is for 'the most outstanding yacht of the year in RORC races'.

SOMERSET MEMORIAL TROPHY WINNERS

Year	Winner	Year	Winner
1966	*Fanfare* (G. P. Pattinson)	1971	*Morning Cloud* (Rt Hon. E. Heath)
1967	*Pen Duick III* (E. Tabarly, France)	1972	*Prospect of Whitby* (A. Slater)
1968	*Phantom* (G. P. Pattinson)	1973	*Frigate* (R. A. Aisher)
1969	*Prospect of Whitby* (A. Slater)	1974	*Keoloha* (Mr and Mrs L. Holliday)
1970	*Noryema VII* (R. Amey)		

The world's first inshore/offshore series started in 1957 was the Admiral's Cup (organized by the RORC). That is a series of races comprising one or more races round buoys during the day and one or more ocean races extending several days and nights: such events have increased in popularity with the suitability of the modern ocean-racing yacht to compete in both types of race. It attracts the most competing nations, the greatest so far being 17 in 1971. There is a maximum of three boats in each national team. The greatest number of yachts to compete was 48 in 1973. (In 1971 several teams were incomplete.) The courses are the Channel race, two inshore races in the Solent and the Fastnet race. In the Channel race and Fastnet, the yachts also compete against individual entries but have their own separate starts. A points system on all four races finds the winning team.

The origin was of an informal nature. The cup was presented by several RORC members including the club's then Admiral, Sir Myles Wyatt (hence the name), to be competed for by British and American yachts, when the latter came to England for the Fastnet race and Cowes Week. The inshore races were mixed into events with other yachts until 1971 when the Admiral's Cup yachts were given separate inshore courses in the Solent. The yachts race on current rating rules and time allowance: the upper and

Start of an Admiral's Cup race off Cowes.

The Admiral's Cup.

ADMIRAL'S CUP PARTICIPATION

	1957	1959	1961	1963	1965	1967	1969	1971	1973	1975
Argentina							●	●	●	●
Australia					●	★	●	●	●	●
Austria								●		
Belgium								●	●	●
Bermuda							●	●	●	
Brazil								●	●	●
Britain	★	★	●	★	★	●	●	★	●	★
Canada										●
Denmark									●	
Finland						●	●		●	
France		●	●	●	●	●	●	●	●	●
Germany				●	●	●	●	●	●	●
Holland		●	●	●	●	●	●	●	●	●
Hong Kong										●
Ireland					●	●		●	●	●
Italy							●	●	●	
New Zealand								●		
Norway										●
Poland								●		
Portugal									●	
USA	●		★	●	●	●	★	●	●	●
South Africa								●	●	●
Spain						●	●			●
Sweden			●	●	●			●		●
Switzerland										●
Total Teams	2	3	5	6	8	9	11	17	[illegible]	19

★ Winners ● Other competing teams

lower limits of rating (and therefore size) have been changed from time to time. In 1975 the limits were between IOR 30 and 44 ft, which means that in effect no yacht is larger than about LOA 55 ft. The largest yacht ever to have competed in an Admiral's Cup team was *Iorana* (Wolfgang Denzil, Austria), LOA 63 ft, in 1971. Because of subsequent upper size limits, this cannot now be exceeded.

Competitions which have followed the Admiral's Cup idea and are now held in other parts of the world include the Onion Patch Trophy in the USA which includes the Newport–Bermuda race; the Southern Cross Cup which includes the Sydney–Hobart race and the Rio Circuit in Brazil. The Ton Cups also adopted the scheme of the inshore/offshore series pioneered by the Admiral's Cup. The progress of the Admiral's Cup is shown in the table opposite:

Transatlantic races for ocean-racing yachts are irregular and have had varying points of departure (usually a port in the USA or from Bermuda) and arrival (usually in England, Spain or Sweden). Various clubs have taken part in the organization, but frequently the Cruising Club of America in conjunction with European clubs. All races have been from west to east (except for single-handed ocean racing, p. 168). From 1866 to 1975 there have been 20 transatlantic races – 14 of these have been since 1945: in 1950, 1951, 1952, 1955, 1956, 1957, 1960, 1963, 1966, 1968, 1969, 1972, 1974 and 1975. The longest in distance was the race in 1951 from Havana, Cuba to San Sebastian, Spain which had the smallest number of starters, 4, since 1887. The shortest course is from Newport, R.I. to Cork, Ireland, 2668 miles. This latter course gave the fastest time on any transatlantic race since 1905: 12 days 5 hours 43 minutes in 1969 by the yawl *Kialoa II* (John Kilroy, USA), LOA 73 ft. This race marked the 250th anniversary of the Royal Cork Yacht Club, the oldest yacht club in the world.

The [illegible] *Atlantic* gained the [illegible] for the fastest time [illegible] transatlantic [illegible] in 1905.

The smallest yacht to win a transatlantic race was the Royal Naval Sailing Association *Samuel Pepys*, LOA 30 ft 9 in, skippered by Commander Erroll Bruce, RN. This was in 1952 from Bermuda to Plymouth. The biggest yacht ever to win was the 185 ft schooner *Atlantic* (Wilson Marshall, USA) racing from Sandy Hook to the Lizard in 1905. She also has the record for the fastest time in any transatlantic race at 12 days 4 hours 1 minute 19 seconds (10·32 knots). The largest number of starters was 42, in the 1966 race from Bermuda to the Skaw, Denmark.

The first transatlantic race was also the most tragic. Three yachts took part: *Henrietta*, 107 ft, James Gordon Bennett; *Fleetwing*, 106 ft, George and Franklin Osgood; *Vesta*, 105 ft, Pierre Lorillard. All three were from the east coast of the United States. The race began on 11 December 1866 off Staten Island, New York. *Henrietta* won, finishing at the Needles, Isle of Wight on Christmas Day in a time of 13 days 21 hours 45 minutes. This was the only transatlantic race sailed in winter. The tragic event was on *Fleetwing* from which on 19 December in a gale, eight men were washed overboard at night. Two managed to get back on board, but the six were not recovered.

In Great Britain the number of races for IOR, cruising yachts and habitable boats generally is no longer listed by the national authority owing to the great number of varying events by different clubs and organizations all round the coast. However, the RYA shows as many races as practicable in its annual fixture list: in 1974 this was 1521 for all types of sailing-boat. Power-boat fixtures are additional. Most of the more prestigious offshore races, but not all, are given by organizations. These are the Royal Ocean Racing Club (giving 20 races in 1974), the Solent Points Championship (12), Junior Offshore Group (13), the East Anglian Offshore Association (12), the North East Cruiser Racing Association (12), the Irish Sea Offshore Racing Association (17) and the Clyde Cruising Club (12).

Two-man race

A new type of ocean race was started in 1966 in which the main limitation was that the crew consist of two men only. This was the Round Britain race thought out by Colonel H. G. Hasler. The race was to begin and end at Plymouth and there were 48-hour compulsory stops at Crosshaven, Castlebay, Lerwick and Lowestoft. The main prize was for the first to finish, but there was a subsidiary prize on time allowance. The race was organized by the Royal Western Yacht Club of England and sponsored by two newspapers. Winner of the first race from 16 other competitors was the 42 ft trimaran *Toria*, sailed by Derek Kelsall and Martin Minter-Kemp. The multihull yacht covered the 2079 miles in 11 days 17 hours 23 minutes – stopovers not included. The race, held again in 1970, was won by the largest competitor, the 71 ft *Ocean Spirit*: this showed that such a size of boat could be handled by two men (Robin Knox-Johnston and Leslie Williams).

The greatest number of starters was in the 1974 race (the event is scheduled for four-year intervals and sponsored by *The Observer*) when 66 crossed the Plymouth starting-line – 43 single-hulled yachts and 23 multihulls. The record for the course was the same year when *British Oxygen*, a 70 ft catamaran sailed by Robin Knox-Johnston and Gerry Boxall, won the race in 10 days 4 hours 26 minutes (8·50 knots). The smallest yacht to complete the course was the Quarto class sloop *Petit Suisse* (Switzerland, Beat Güttinger and Albert Schiess), in 17 days 20 hours 28 minutes (5·05 knots).

Two multihulls have capsized on the course, the catamaran *Apache Sundancer* in 1970 and *Triple Arrow* in 1974, which was eventually righted and sailed on. The type of course and manning rules make this race unique.

Just one of the many trophies available to IOR classes. This is the Captain James Cook Trophy presented to the British Junior Offshore Group by its Australian counterpart for team racing in small yachts. Britain, France, Australia compete every three years.

Fireball: designed in Britain, popular world-wide. The scow-type hull is clear here

Mirror dinghy; the world's numerically largest two-man dinghy class

Laser: fastest-growing class of single-handed dinghy

British Oxygen, 70 ft catamaran, holder of the Round Britain race record.

Right
Triple Arrow, sailed by Brian Cook, which capsized in the 1974 Round Britain race.

The race which has established itself as attracting the largest number of starters annually in Britain is the Round the Island race of the Island Sailing Club held in late June or early July. The course begins and ends at Cowes and is round the Isle of Wight.

The race was first held in 1926, the principal award being a Gold Roman Bowl, though there are numerous other subsidiary prizes. (There are other races which cover the same or very similar course, but these do not attract the interest of the Island Sailing Club event.) Since 1945 entries

Two-girl crew who completed the 1974 Round Britain race, Clare Francis (at tiller) and Eve Bonham.

The trimaran *Three Cheers* holds the record time for the Round the Island race. She was sailed by Mike McMullen and designed by Dick Newick.

Right
Starters in the Round the Island race. They are impossible to count as they emerge from the mist and sail towards the western tip of the Isle of Wight.

have increased yearly. The Round the Island race has the biggest single entry of any race in Europe. This entry was a record in 1974 with a total of 479 starters (there were 516 entries). Of these, 405 starters were in the IOR class and the remainder were cruising one-designs and multihulls. The fastest time ever taken on the race was 5 hours 29 minutes 30 seconds in 1973 by the trimaran *Three Cheers*, sailed by Mike McMullen, over the nominal distance of 60 miles. Tidal streams ensure that less than this distance is not sailed through the water. The speed was, therefore, a little less than 10·9 knots.

The fastest time for a single-hulled yacht is 6 hours 11 minutes by the 62 ft Australian *Apollo* (J. Rooklyn) also in 1973 (9·8 knots nominal).

The only person to win the Gold Roman Bowl three times running was the Rt Hon. Edward Heath, when he was Prime Minister, in 1971, 1972 (*Morning Cloud* (II)) and 1973. In 1973 *Morning Cloud* (III) was a yacht of the same name but a different craft from the one to win in the earlier years.

The earliest time that the race has been started was in 1973 when the first gun (10 minute warning) for the first class to leave was at 04.50 BST.

Cowes Week

The most important regatta week for yacht racing in Britain is Cowes Week. It is generally considered the pre-eminent regatta of the world. It is held annually in the first week of August. The first Cowes Week race ever began at 09.30 on 10 August 1826 for a £100 gold cup under the flag of the Royal Yacht Club (later the RYS). There was another race the next day, and there was a ball and a firework display. Over the years the classes, clubs and races multiplied. By 1845, a contemporary report mentioned 'Cowes presented a most magnificent scene, upwards of 80 vessels being under sail.' Then in 1894 it was reported 'The Cowes Week has always been an assemblage of aristocrats, but the year 1894 has eclipsed all previous gatherings. Never have so many yachts graced the beautiful waters of the Solent. A cloud of craft was anchored in Osborne Bay, up the Medina, and right away to Gurnards Bay, and flew every national ensign, American, French and German predominating.'

Noryema, owned by Ron Amey of Britain, is the only foreign yacht to have won the Bermuda race.

The importance of Cowes Week comes from its tradition, its continuing royal patronage from the days of King William IV, its regular good breezes and exacting tidal courses, the waterfront which provides a grandstand for racing, the ability of the River Medina to accommodate great numbers of yachts and its social aspect in the English 'season'.

Until 1914 the racing was chiefly for the big cutters and small raters; from 1919 to the 1950s the 'metre' boats and one-designs were considered most important, but handicap classes and ocean racers kept increasing. Now the ocean-racing boats predominate with one-designs being only local in character. Prince Philip usually attends in HMY *Britannia* with other members of the royal family. Cowes Week is now always nine days: Saturday to the Sunday of the following week-end. It is now linked to ocean racing because the Channel race is over the first week-end and the Fastnet (or another RORC race on even years) begins on the last Saturday. The Admiral's Cup takes place during the week on odd years. The Royal Yacht Squadron has combined with other clubs in organizing the week as the 'Cowes Combined Clubs'. Balls and a firework display still occur. Specialization had meant such features as rowing and races for dinghies have moved away to their own regattas over the years.

A record number of yachts took part in the 1973 Cowes Week as follows:

IOR yachts (which were divided into four classes and varied from 24 ft to 73 ft in length): 394.
One-design classes (of which there were eleven classes up to about 27 ft long): 297.

American racing

The first offshore race in the modern way (that is for yachts of moderate size – not over 70 ft manned largely by amateurs) was on the east coast of the USA in 1904. It was from Brooklyn to Marblehead passing outside Long Island, a distance of 330 miles. The race was organized by Thomas Fleming Day, Editor of the magazine *Rudder*. Six yachts took part, all less than 30 ft waterline and the winner was *Little Rhody* owned by Charles F. Tillinghast.

The oldest ocean race that is still in existence is the Bermuda race. It was first sailed in 1906 from Gravesend Bay, Brooklyn to Bermuda, 660 miles. Three yachts took part. The winner was *Tamerlane*, 38 ft 3 in, owned by Frank Maier and sailed by Day. Her time was 5 days 6 hours 9 minutes. In 1907 12 yachts started in the race. One of the competitors *Zena*, from Bermuda, was the first yacht ever to engage in ocean racing with a Bermuda rig. Three more races to Bermuda took place before the First World War, in 1908 from Marblehead, in 1909 and in 1910 but with decreasing entries. The last race consisted of two schooners.

In 1923 the race to Bermuda was revived and started from New London, Connecticut. It was organized by the Cruising Club of America, which had been founded in the previous year. Twenty-three yachts took part. The race was held again in 1924, but there were only 14 starters out of which only 4 had taken part in 1923, and thereafter the race was biennial on even years: thus it alternated with the Fastnet race. The pre-war series continued until 1938. In 1936 the course was changed again, so that the start was from Newport, R.I. This course remains in force today. The Newport–Bermuda race has been held every even year from 1946. Until 1970, the rating rule of the Cruising Club of America was used: from 1972 the

International Offshore Rule was used. The length of the race is 635 miles. The largest number of yachts to start in the Bermuda race was 167 in June 1974.

The fastest time on the course is held by *Ondine*, 79 ft ketch owned by S. A. 'Huey' Long, which in 1974 had an elapsed time of 2 days 20 hours 8 minutes 22 seconds (9·32 knots). (She was awarded a 16 minute bonus on this because of going to check a cruising yacht suspected of being in distress, but this cannot be included in the record.) The previous record was by *Bolero* in 1956, and the pre-war record was 2 days 23 hours 35 minutes 43 seconds by *Highland Light* (61 ft 8 in) in 1932.

The only foreign yacht to have won the Bermuda race was the British yacht *Noryema*, 48 ft, owned by Ron Amey and sailed by Ted Hicks in 1972.

The Onion Patch Trophy is on the lines of the Admiral's Cup being for three-boat national teams in an inshore/offshore series. The long race of the series is the Newport–Bermuda race. The series has never attracted more than five nations, for geographical reasons. The USA has always won the trophy since its inception in 1964, except in 1966 when it was won by Britain.

The most outstanding rescue in the history of the Bermuda race was in 1932. The British cutter *Jolie Brise*, skippered by Robert Somerset, rescued the crew of another competitor, the American schooner *Adriana*, on fire at sea on the first night out on the race, 26 June 1932. Somerset took his yacht across the stern of the burning yacht, when ten members of the crew were able to jump on to the deck of *Jolie Brise*. Only the helmsman, Clarence Kosley, who had stayed at the wheel until last, failed to reach *Jolie Brise*. He fell between the two yachts and could not be found afterwards. *Jolie Brise* abandoned the race to take the rescued men to port. Later Somerset was awarded the Blue Water Medal of the Cruising Club of America and was presented with a gold watch by the President of the United States.

Somerset, who held the DSO and was heir to the Duke of Beaufort, was drowned on 28 February 1965. When trying to enter the harbour of Mandraki, Rhodes, in the Mediterranean, his yacht *Trenchemer* (a 72 ft

The 79 ft *Ondine*, Bermuda Race record holder. Yes, she is all one boat: the ketch rig is unorthodox.

Right
Start of the Bermuda race from Newport, Rhode Island.

Windward Passage, 73 ft, averaged 10·02 knots on the Los Angeles to Honolulu race.

Right
The American yacht *Dynamite*, was first winner of the Canada's Cup using the International Offshore Rule.

steel 1932-built Sparkman and Stephens design and the first to be built in England to the RORC rating rule) hit the rocks. Somerset, true to character, went to try and rescue two women who were down below, but both he and the women were lost. A friend, William Johnston, and three Spanish crew were saved.

Pacific

The world's regular ocean-racing fixture with the longest distance is the trans-Pacific race. The first one started 11 June 1906 and three yachts, all very large, took part. The winner was *Lurline*, 86 ft 3 in, owned by H. H. Sinclair. The race is of a special nature because it is held in the prevailing trade winds and is one of the very few events where it is certain that the wind will blow fair throughout. For this reason it has always been difficult to find a suitable system of rating and time allowance: most other systems assume that a yacht will spend a mixture of most races, beating and reaching as well as running. In the 1975 'transpac' the International Offshore Rule was used, but parts of the formula were altered to compensate for the expected winds from astern for almost all the race.

Races were held from US west coast ports, usually San Pedro to Hawaii in 1908, 1910, 1912, 1923 and 1925. The maximum number of starts in any of these races was four!

Canada's Cup, first raced in 1896.

There were eight more races between 1926 and 1941. From 1947, the race has been held biennially by the Transpacific Yacht Club on odd years starting at Los Angeles and finishing at Honolulu, a distance of 2225 miles.

The largest number of yachts to start in the race was 72 in 1969. The record time for the course was sailed by *Windward Passage*, 73 ft (Mark

Johnson), in 1971 in a time of 9 days 5 hours 34 minutes 22 seconds (10·02 knots). *Windward Passage* had a speed of over 20 knots (the speedometer upper reading) at times. On another race in the same year from Miami to Montego Bay, Jamaica, the same yacht reported a noon to noon run of 262 miles (10·9 knots).

Canada's Cup

The oldest match race sailed between the United States and Canada, which is still competed for, is Canada's Cup. The trophy was put up in 1895 by the city of Toledo, Ohio and the first race for it was in 1896 when the Lincoln Park Yacht Club of Chicago represented by *Vencedor* lost to *Canada*, a yacht of the Royal Canadian Yacht Club of Toronto. This first race was under a time-allowance system using the Seawanhaka rating rule of that time. The races are always held at a venue on the Great Lakes. Twelve matches have been held between 1896 and 1972. In these the Royal Canadian Yacht Club has won four times and an American club (which has varied) won the others. Until 1907, the series was under the current Girth and then Universal rule and the best of three races decided the winner. In 1930, 1932, 1934 and 1954, the races, still the best of three, were held in International 8-metres.

In 1969, the series became an inshore/offshore event for yachts of 37 ft rating Cruising Club of America rule. In 1972, *Dynamite* (USA) beat *Mirage* (Canada) using a fixed rating of 32 ft under the International Offshore Rule. Canada's Cup is the only inshore/offshore match race with one race between 150 and 250 miles and two 21-mile Olympic courses used for the short events. The race was held in September 1975 when the US *Golden Dazy* won.

Largest numbers

The largest number of starters in any offshore race is in the annual Newport Beach (near Los Angeles) to Ensenada, Mexico event. The distance is 130 miles. In 1973 this race recorded the highest number of yachts ever to start in a single race: 573. The sponsors of the race are the Newport Ocean Sailing Association and the race is held on the week-end closest to 5 May. In the 1974 race, the 44 ft catamaran *Seabird* was first to finish of 563 starters in 17 hours 1 minute (7·32 knots).

The race of any sort to attract the most entries and starters in the world, though not an offshore race because much of it is in protected waters is the annual Round Sjaelland race. The course of 275 miles is round the island on which Copenhagen is situated and is organized by the Royal Danish Yacht Club. The race is held as near to midsummer's day as possible. All the yachts are keel boats and open one-designs do not compete. Many of them are small locally handicapped cabin yachts which are common in Scandinavia. In 1974, 942 yachts entered of which 139 were IOR class boats.

Winner of the second Canada's Cup series in 1899. The 44 ft 10 in American sloop *Genesee*. She was the first American winner of the trophy, which was most recently sailed for in 1975.

SORC

The Southern Ocean Racing Conference is a series of races held annually in February and March off the coast of Florida. Every year it is the first of the season's offshore series of international standard in the world. The series is unique in extending over a six-week period with breaks between races. The races are St Petersburg to Anclote Key and back (101 miles), St Petersburg to Fort Lauderdale (400 miles), Miami and return on an 'ocean triangle' (132 miles), the Lipton Cup, a 38 mile course from Miami to Fort Lauderdale, the Miami-Nassau race, 176 miles and the Nassau Cup a short course.

The series is outstanding in attracting new boats because of the time of year and entries with crews of designers, builders and sailmakers seeking practical success for their latest creations.

The series began in 1941 under the name of 'The Winter Circuit', though of the individual events, the Lipton Cup began in 1928 and the Miami–Nassau in 1934. The first boat to win the circuit was the famous Sparkman and Stephens yawl *Stormy Weather* (see Fastnet results) though this 1941 win was the only tie in the series with *Gulf Stream*, 70 ft: nowadays points systems are arranged to avoid this, sometimes by specifying an overriding race. *Stormy Weather* when owned first by Robert W. Johnson and then by William Labrot won the Nassau race five times in a row, an unbeaten record. It also shows the success of this design long after she had been acclaimed in England. Amazingly *Stormy Weather* won the circuit again in 1948.

Carleton Mitchell

The only man to win the circuit three times was owner/skipper Carleton Mitchell, first in the 1937 Rhodes-designed *Caribbee*, 57 ft 6 in, in 1952 and 1953 and in the Sparkman and Stephens-designed 38 ft 8 in *Finisterre* in 1956. Both were beamy centreboarders to the CCA Rule. The most outstanding rule cheaters to have success in the series were *Hoot Mon*, 39 ft light displacement boat to the CCA Rule which resembled a Star class one-design and was sailed by sailors of that class – Worth Brown, Charles Ulmer and Lockwood Pirie; winning in 1954 and 1955 and the other rule cheater was to the IOR in 1973, when the 'cat-rigged ketch' *Cascade*, 38 ft with two 'mainsails' and no forward headsails, won three of the races but not the series.

Fastnet race winner, *Stormy Weather*, was also in 1934 first-ever winner of the now well-established SORC.

Caribbee, 57 ft 6 in.

Right
Conquistador, a Cal 40 which won the SORC. The Cal 40 class was a leading influence in ending of the long keel for racing yachts in the early 1960s.

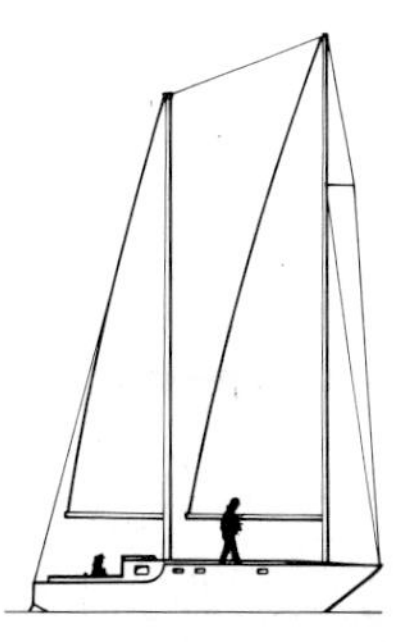

Profile of the 'cat-rigged ketch', *Cascade*.

WINNERS OF THE SORC SERIES

Year	Winner
1941	Tie, *Stormy Weather*, William Labrot, and *Gulf Stream*, Dudley Sharp
1947	*Ciclon*, A. Gomez-Mena and M. Bustamente
1948	*Stormy Weather*, Fred Temple
1949	*Tiny Teal*, Palmer Langdon and Richard Bertram
1950	*Windigo*, Walter Gubelmann
1951	*Belle of the West*, Will Erwin
1952	*Caribbee*, Carleton Mitchell
1953	*Caribbee*, Carleton Mitchell
1954	*Hoot Mon*, Brown, Pirie and Ulmer
1955	*Hoot Mon*, Brown, Pirie and Ulmer
1956	*Finisterre*, Carleton Mitchell
1957	*Criollo*, Luis Vidana
1958	*Ça Va*, J. W. Hershey and Robert Mosbacher Jr
1959	*Callooh*, Jack Brown and Bus Mosbacher Jr
1960	*Solution*, Thor Ramsing
1961	*Paper Tiger*, Jack Powell
1962	*Paper Tiger*, Jack Powell
1963	*Doubloon*, Joe Byars
1964	*Conquistador*, Fuller E. Callaway III
1965	*Figaro IV*, William Snaith
1966	*Vamp X*, Ted Turner
1967	*Guinevere*, Geo. Moffett
1968	*Red Jacket*, Perry Connolly
1969	*Salty Tiger*, Jack Powell and Wally Frank
1970	*American Eagle*, Ted Turner
1971	*Running Tide*, Jakob Isbrandtsen
1972	*Condor*, Hill Blackett
1973	*Muñequita*, Jack Valley and Click Shreck
1974	*Robin Too II*, Ted Hood
1975	*Stinger*, Denis Connor

Start of the Sydney–Hobart race.

Helsal, only ferro-cement yacht to win a major ocean race and fastest time over the Sydney–Hobart course.

The first GRP boat to win the Southern Ocean Racing Conference as it became known in 1953 was *Paper Tiger*, 40 ft, designed by Charley Morgan of Fort Lauderdale for Jack Powell. The plastic boat won in 1961 and 1962 and had a major influence in the USA in the adoption of GRP for racing yachts which were to be sailed hard. In 1969 Powell won again in *Salty Tiger* designed by Bob Derecktor. The first winner to have a separate fin and skeg rudder was *Conquistador*, 40 ft Cal 40 designed by Bill Lapworth and owned by Fuller E. Callaway in 1964. The 1966 winner *Vamp X* was also a Cal 40, owned by Ted Turner who was to achieve greater fame as an offshore skipper/owner in *American Eagle* and *Lightning*. These boats (and other designs with separate rudders, see Section 5) spelt the end of the 'long keel'. All new sailing-yachts with any pretensions of speed have had separate rudders aft since 1966. The International Offshore Rule has replaced the CCA Rule in the series since 1971.

The designers to have had more winners than any other are Sparkman and Stephens of New York.

Each year there are numerous offshore races in North America, east coast, west coast and Great Lakes. In the last the 333-mile Chicago to Mackinac race across the length of Lake Michigan has been sailed without interruption every year since 1921, making it the offshore race with the longest annual run: being on the Lakes the Second World War did not prevent it. It was also sailed 14 times before 1916. Other typical American courses are: round Block Island, Swiftsure race (in the Pacific north-west), San Francisco–Newport, Galveston–Vera Cruz, Marblehead–Halifax, Bayview–Mackinac, Sandy Hook–Chesapeake Bay, Cape May–Brenton Reef.

Australian races

The most important annual ocean race in the Southern Hemisphere is the Sydney to Hobart race, distance 630 miles. The race was first held in 1945 at the instigation of Captain J. H. Illingworth, RN. It invariably starts on 26 December, Boxing Day and being a public holiday in midsummer attracts the largest crowds in the world that regularly turn out to see an ocean race begin. The race has a tough reputation, because the course has to cross the Tasman Sea where strong westerlies prevail. The most yachts to start in one race was 92 in 1973. The course record is held by the first yacht to finish in the 1973 race, the 72 ft 4 in *Helsal* (Dr Tony Fisher, Australia) in the time of 3 days 1 hour 32 minutes (8·57 knots). The yacht was unusual among racing boats in being constructed of ferro-cement, though this is common enough in cruising boats of all sizes. Her displacement was 90 000 lb: heavy for a record-breaker.

John Illingworth

The Southern Cross is the Australian contest comparable with the Admiral's Cup, being an inshore/offshore series. There is one offshore race of 180 miles, two inshore events and then the long race of the series for three-boat national teams is the Sydney–Hobart. The series is held every odd year: the British team has won it once – in 1973. (The series actually finishing in the first week of January 1974.) Two English yachts have been over-all winners of the Sydney–Hobart: Captain J. H. Illingworth's *Rani* in the 1945 inaugural race and *Morning Cloud*, 34 ft, owned and skippered by the Rt Hon. Edward Heath, then Leader of the Opposition, in 1969. *Morning Cloud* was not part of the British Southern Cross team competing in the same race.

Ceil III, winner of the Sydney–Hobart race in 1974.

Wing-sail development by 1974 on C class catamaran, *Helios*, in Australia.

Multihulls

The only regular established offshore race in Europe for cruising multihulls is the annual Crystal Trophy, organized by the RYA with help from local clubs. The course is from Cowes to CH1, off Cherbourg, to Wolf Rock Lighthouse, finishing at Plymouth, a distance of 311 miles. The maximum number of starters was 28 in 1974. The race was first held in 1967 and the winners have been as follows:

1967	*Tomahawk*	(K. Isted and G. Tinley)
1968	*Tomahawk*	(K. Isted and G. Tinley)
1969	*Wardance*	(G. Tinley)
1970	*Trifle*	(Major-General R. Farrant)
1971	*Minnataree*	(G. Boxall)
1972	*Minnataree*	(G. Boxall)
1973	*Silmaril*	(D. Walsh)
1974	*Rage* (FT)	(D. Palmer)

Results were determined by the Portsmouth Yardstick. The record time for the course was by *Trifle* in 1968. She took 1 day 17 hours 6 minutes (7·57 knots). She was first to finish the race every year from 1968 to 1972. Built in 1967, she is a 42 ft trimaran, 27 ft beam, designed and built in foam sandwich construction by Derek Kelsall. In May 1971, *Trifle* sailed from Poole to Cherbourg Harbour entrance (62 miles) in exactly 6 hours (10·33 knots). She had recorded a 500 metre record attempt at 16·7 knots in 1972, and this is an interesting comparison between average offshore and 'bursts'.

The most significant mishap in the race was the capsize of the Rudy Choy catamaran *Golden Cockerel* in 1967, east of the Isle of Wight, after sailing only 15 miles. There were no casualties and the catamaran was salvaged, but it resulted in revised sailing techniques and new safety precautions such as improved access to the life-raft from under the hull.

With certain exceptions, the regular offshore racing of multihull craft is not indulged in. The exceptions are the multihull class in the Round the Island race, and the four-yearly Round Britain. The Multihull Offshore Cruising and Racing Association (MOCRA) exists to encourage offshore competition in multihulls.

Golden Cockerel, which capsized on the Crystal Trophy race in 1967 and remained bottom-up.

Trifle, course record-holder on the Crystal Trophy multihull race.

Right
C class catamarans in the Little America's Cup. In the 1967 contest *Lady Helmsman* (Britain) beat *Quest III* (Australia).

Longest

The longest offshore race for multihulls to be held on a regular basis is the Multihull Transpac from Los Angeles to Honolulu, which is biennial. The fastest time recorded for this was in the race of July 1974, when the 45 ft catamaran *Sea Bird*, owned by Bob Hanel, Cabrillo Beach Yacht Club, Los Angeles completed the 2225-mile course in 8 days 19 hours 31 minutes 38 seconds (10·53 knots). *Sea Bird* was the lightest boat for her length ever to have competed in this event. The Transpac is traditionally a downwind race and the yachts are developed in the expectation of these conditions. However, in this fastest race there was 1200 miles of close-hauled and reaching weather before the wind came aft to give the fastest speeds: a maximum speed of 28 knots in one burst was reported by those on board. The fleet in 1974 consisted of four catamarans and two trimarans.

Little America's Cup

Indisputably the foremost international match racing in catamarans is the International Catamaran Challenge or 'Little America's Cup'. In 1959/61 the American magazine *Yachting* held a One-of-a-kind regatta for dinghy-type cats which was won by *Tiger Cat* to the international C class (maximum length 25 ft, maximum beam 14 ft and carrying up to 300 ft^2 of sail with a crew of two, but with few other restrictions).

Meanwhile in Britain a one-of-a-kind of the same type had been won by *Hellcat* designed by Rod Macalpine Downie and sailed by John Fisk. *Hellcat* was reckoned to be faster than *Tiger Cat* and a challenge by the Chapman Sands Sailing Club with her was accepted for 1961 by the Eastern Multihull Sailing Association of Long Island Sound. *Hellcat* was of glass fibre with only 12 ft beam which was 2 ft under maximum allowed, twin daggerboards of laminated agba and a fully battened mainsail and small jib supported by an alloy spar. In the event the American defender was *Wildcat* sailed by John Beery of San Francisco. She lost four races out of the five for the 'Little America's Cup' sailed in Long Island Sound.

The trophy stayed in Britain with victories in 1962 when *Hellcat* beat *Beverly* (USA), in 1963 when *Hellcat II* beat *Quest I* (Australia), in 1964 when *Emma Hamilton* beat *Sea Lion* (USA), in 1965 when *Emma Hamilton* beat *Quest II* (Australia), in 1966 when *Lady Helmsman* beat *Game Cock*

(USA), in 1967 when *Lady Helmsman* beat *Quest III* (Australia), in 1968 when *Lady Helmsman* beat *Yankee Flier* (USA). It really began to look like the America's Cup in reverse, until in 1969 two young Danes, Gert Frederiksen and Leif Wagner arrived at Thorpe Bay on the Thames, the site of the previous defences, with *Opus III* to challenge Reg White and John Osborne in *Ocelot*. The Danes won three races and the British won three races and then in the final decisive race the Danes won and took the trophy from what was beginning to look like a British monopoly. In 1970 the Australian *Quest III* beat the Danish *Sleipner* by four races to three. Since then the trophy has remained in Australia, but it is still valid for a challenge from a single C class catamaran backed by a club of any country.

Lady Helmsman – 25 ft, beam 14 ft, weight 600 lb – designed by Rod Macalpine Downie with wing mast 36 ft high rigged by Austin Farrar is now preserved in the National Maritime Museum, Greenwich.

Inshore Trophies

Some of the old inshore trophies, such as the One Ton Cup and Canada's Cup, are now seen to have been altered to offshore types of event to cater for the modern ocean-racing yacht. In each country prominent in sailing, there are a number of regular trophies for certain classes or sorts of race, which remain for inshore class yachts. In other words, the course or courses of the event would be in comparatively sheltered water, raced round

The amazing American 6-metre class boat *Goose* designed by Stephens in 1938, which has been continuously modernized and now sails at Seattle.

Right
6-metre yachts have continued to sail in the Swiss Lakes, while salt-water sailors have abandoned them in favour of habitable boats.

Left
British-American Cup races in 1951.

triangles, Olympic courses or some irregular course round buoys and take between 2 and 6 hours. Keel boats would be used, that is without accommodation, deck structures or motors (for example Dragons, Solings). There are some particularly famous trophies of international standing and some facts about these now follow. Events for centreboard boats and sailing-dinghies are listed later.

British-American Cup

The British-American Cup was instituted in 1920, when international racing was not as common as it is today. It was the idea of the American Paul Hammond and was to encourage the use of a common class, the 6-metre class, which had not previously been raced in the USA. The rules ensured that races were alternately in Britain and America regardless of who won them (unlike the America's Cup rules). The first races with a team of four 6-metres on each side were in the Solent in July and August 1921. The American team was organized by the Seawanhaka Corinthian Yacht Club of Oyster Bay. The British won this first match and there was a return match in 1922 in Oyster Bay won by the Seawanhaka. Because of the great interest these races caused, the Seawanhaka Cup was in 1922 turned over to the 6-metre class (and immediately won by the Royal Northern Yacht Club of the Clyde). Six-metre racing was considered the highest standard in yacht racing between the wars. Ocean racing had not caught on in the early 1920s and yet the big racing cutters were nearly dead. The expression 'like 6-metre racing' is still taken to mean very close racing between equally matched boats. The American historian of the Seawanhaka Corinthian Yacht Club commented 40 years later about the 6-metre yachts: 'The club's acceptance of this great foreign class had resulted in a decade of international racing never previously equalled in United States yachting history.'

The British-American Cup matches continued up to 1938, the American team asserting their superiority by winning all matches after 1928. In 1938 in Long Island they beat the British, the outstanding American yacht being *Goose* (George Nichols) designed by Sparkman and Stephens. The best British boat was *Circe* designed by David Boyd which went on to win the Seawanhaka Cup and then defended it in 1939 on the Clyde. (This had consequences 18 years later, when the design of a British 12-metre to challenge for the America's Cup was being sought.) The dimensions of *Goose* were LOA 37 ft, LWL 23 ft 8 in, beam 6 ft, draught 5 ft 5 in and sail area 474 ft^2.

After the Second World War, the cup (or rather cups, as there were successive ones to be won outright after a certain number of series wins) was raced for by the depleting fleet of 6-metres in 1949, 1951, 1953 and 1955. *Goose* took part in all these matches.

Six-metre yachts were not raced in Britain after 1956. Interest, however, persisted in pockets round the world, including Scandinavia and the Swiss Lakes. The real centre has become Seattle, Washington (State), USA. There, the stimulus has been the Australian-American Cup for 6-metres. A single Australian, John Taylor, built three boats in successive challenges between 1968 and 1973. The nearly historic *Goose* from 1938 beat the first challenger *Toogooloowoo IV* in 1968. The latter became *St Francis VI*, owned by a San Francisco syndicate which beat the next challenger, *Toogooloowoo V*. The third challenger, *Pacemaker*, designed by Sparkman and Stephens also failed and she was defeated by *St Francis V* designed by Gary Mull (USA). *Goose* remained in Seattle and was heavily modified in hull shape to bring her up to date. The experience in the Sparkman and Stephens 12-metre America's Cup defenders led to these alterations. In Puget Sound where the 6-metres sail the newer challengers and defenders are joined in week-end racing by 25 older American boats and 8 Canadian. In 1976 another Australian-American Cup for match racing in 6-metres was due to take place in Sydney. (See also p. 193.)

Meanwhile the British-American Cup was revived in 1974, using Olympic Soling class one-design keel boats. In the first series, on the Clyde, the Royal Northern Yacht Club beat the Seawanhaka Corinthian Yacht Club.

Ethelwynn, first defender of the Seawanhaka Cup owned by C. J. Field, designed and sailed by W. P. Stephens. Waterline length was 15 ft 2 in and the boat cost $615.78.

Seawanhaka

The Seawanhaka Cup was, as mentioned, used for match racing in 6-metres after 1922, but was an earlier international trophy. Foreign yacht clubs could challenge the Seawanhaka boat. Its importance was that it began international racing in a small rated class of about 23 ft at a time when much larger yachts had the prestigious cups. An English challenger *Spruce IV*, 23 ft 3 in (J. A. Brand), a half-rater from the Minima Yacht Club, first raced *Ethelwynn* (W. P. Stephens) in 1895. The American boat won but the Royal St Lawrence Yacht Club of Montreal won the cup in 1896 and then held it until 1905. Owing to defects in the rating rule the boats became an unhealthy low freeboard type with overhangs longer than the waterline length. The Universal Rule (see p. 191) was adopted instead until 6-metres were used for the cup. In 1971 the Seawanhaka Cup was used at Oyster Bay in international elimination match races for Soling class boats when the winner was Bob Mosbacher (USA) who won six straight races over several foreign opponents.

The Scandinavian Gold Cup was used for 6-metres. In 1927 it attracted the largest number of yachts from different countries that had ever raced at one time in the USA. This was seven! In this 1927 series the Swedish yacht *Maybe* (Sven Salen) broke out a genoa, the first ever seen in the USA. *Maybe* won the Gold Cup on this occasion.

The oldest international trophy for centreboard boats still in existence is the New York Canoe Club International Cup first presented in 1884. It was held by the USA against challenges from Britain and Canada on seven occasions up to 1914. Then in 1933 it was won by Uffa Fox sailing *East Anglian* and Roger de Quincy sailing *Valiant* in sailing canoes designed by Uffa Fox. Subsequently Britain won in 1936, 1948, 1959, 1961, the USA won in 1952 and 1955 and Sweden won in 1974.

Coupe de France

The Coupe de France was established on 6 January 1891 by the Yacht Club de France, the senior French club. It was intended to be a kind of America's Cup and was for an international race open to 'any owner of a yacht built in his own country and challenging on behalf of that country'. This rather nationalistic idea was not taken up by any foreign yachtsmen until the Royal Temple Yacht Club of Ramsgate challenged in 1897. From then until 1901 the yachts were 20-tonners under the French measurement rule of the time. In 1898 *Gloria* (Harrison Lambert) (Brit.) steered by Tom Jurd beat at Cannes in inshore races the French defender *Esterel* (Henri Menier and syndicate). The course took about 2½ hours. After this race, *Esterel* was sold by the syndicate to Baron Edmund de Rothschild and the proceeds were used to purchase the One Ton Cup.

Typical early challenger for the Coupe de France, the 10 ton (old French rating rule) Italian yacht *Artica* of 1902.

Right
Coupe de France competitor of 1927, *Aile IV* owned by Madame Virginie Heriot: International 8-metre yacht lost to the British challenger *Siris II*.

From 1902 to 1907 the Coupe de France was sailed in 10-tonners to the French Rule (5 occasions); from 1908 to 1914, 10-metre boats to the International Rule were used (5 occasions); from 1922 to 1949, 8-metre yachts were used (16 occasions). The cup was allocated to the IYRU 5·5-metre class from 1953 to 1967 (13 occasions) and subsequently IOR classes of 5·5-metres (which is the 18 ft rating Quarter Ton Cup level) have competed for the cup in 1971, 1972, 1973 and 1975. Single challenges are, however, no longer favoured in racing, team and points series taking their place. Since 1891 the cup has been contested 48 times and won by France and Britain (15 times each), Switzerland 7, Norway 4, Italy 3, Germany 2, Sweden 1 and Australia 1.

The yacht to win it the most times was the 8-metre *Severn* (Brit.) in 1930, 1931, 1932 and 1933. The last British yacht to win it was the 5·5-metre *Yeoman XV* (Robin Aisher) in 1967. The only Australian yacht to win it was the 5·5-metre *Southern Cross* in 1965.

The Dragon Gold Cup, which is raced for annually by an international fleet, stands out today as being the only remaining established cup event on an international basis raced by an IYRU recognized keel boat. (Of the other existing IYRU keel boats, the Star class is scarcely raced in Europe, while the Solings and Tempests are comparatively new classes.)

The Gold Cup was presented by the Clyde Yacht Clubs' Conference in 1936, after a first international series for Dragon class yachts in Scotland. It has been sailed every year from 1937 (except 1938–46). Seven times it has been sailed on the Clyde, but on the other occasions in Scandinavia and the Baltic. Up to 1973, helmsmen of the following countries have won the cup: Denmark 16 times, Germany 3, Norway 3, Sweden 3, Britain 2, Canada 1, Italy 1, Northern Ireland 1, USA 1.

USYRU

The United States Yacht Racing Union has a number of annual trophies open to yachtsmen throughout Canada, Mexico and the United States. Among the most notable are these: The Congressional Cup is a knock-out match racing event between top yachtsmen to find a national 'Champion among champions'. The series is held annually at the Long Beach Yacht Club, California. The cup was presented in 1965 by the US Congress. The Sears Cup helps to explain the great success of American sailors, because it encourages them from an early age. It is open to crews of three boys or girls aged between 13 and 18, and whose parents are members of a club recognized by USYRU. It dates from 1921 when it was available to Massachusetts only, but since 1951 has been open to clubs throughout North America.

The Clifford C. Mallory Cup is named after the first President of USYRU and is the North American Sailing Championship. The regional associations of USYRU put forward crews who have been found by elimination races. For instance the first crew to win in 1952 was skippered by Cornelius Shields representing YRA of Long Island Sound. In 1971 the Texas Yachting Association was and in 1972 the Florida Sailing Association.

The Prince of Wales Bowl is the North American inter-club championship. It was presented by the then Prince of Wales to the Royal Nova Scotia Yacht Squadron in 1931, but in 1965 the cup became available under the present rules. Various classes of boat are used; for instance, small standard cruiser racers of the same design have been used in recent years.

RYA

The primary national team event run by the Royal Yachting Association is the British club team championship. It includes all of Ireland and is confined to centreboarders. In 1973 there were 12 areas of the British Isles and knock-out competitions were held between all clubs within the areas that entered. Any type of centreboard dinghy with a recognized owners' association may be used in matches of the contest. This championship has the cleverest of all ways of deciding what boat to use. The home club in each match nominates the class of boat to be used. If the away club refuses to race in this class, it has the right to do so, but must then supply all the boats. As this means expense and wear and tear, the system is self-regulating and the RYA is spared any disputes on this score. The team championship originated in the London area only, but became national in 1969. Winning clubs since then have been the University of London Sailing Club (twice), Felixstowe Ferry Sailing Club and West Kirby Sailing Club. The greatest number of teams to compete in a single season was 322 teams in 1971.

The first international championships for women only were held at Quiberon, France in May 1974. These were organized by the French national authority (La Fédération Française de Voile). The object was to interest the IYRU in instituting a class for women only in the Olympic Games in 1980. In these first races, 80 women from 11 countries took part. There were 29 in the single-handed dinghy class (using the Europe class) and 26 boats in the two-man dinghy class (420 class). The former was won by Martine Allix (Fr.) and the latter by Berita Volk and Gonnede de Vos (Neth.).

'The champion racing yachtsman' is impossible to find because the best a helmsman can do is win the championship of his class at a single event or series, even at the Olympic Games. Some of the USYRU eliminations, such as the Congressional Cup, have been mentioned above and these try to find a 'Champion among champions'. In Great Britain, the Endeavour Trophy was established by the Royal Corinthian Yacht Club of Burnham-on-Crouch in 1961. In this the champion of each RYA recognized centreboard dinghy class with more than 200 registered boats sails with his crew. A one-design class is used for the actual races at Burnham and since 1972 the Lark dinghy has been used.

The Endeavour Trophy itself was presented by Beecher Moore and is a model of the 1934 America's Cup challenger *Endeavour* and is intended to perpetuate the memory of the amateur sailors from the Royal Corinthian Yacht Club who manned the J class yacht after the professional crew had struck (see p. 136). The winners of the trophy each year are shown below. The class shown is that from which the helmsman and crew have come from in order to compete. Until 1971 National Enterprise dinghies were used for the races.

Date	**Helm**	**Crew**	**Class**
1961	P. Bateman	K. Musto	Int. Cadet
1962	M. Evans	R. Smith	Int. 12
1963	D. Newman	R. Martin	Nat. Firefly
1964	R. Pitcher	P. Amos	Flying Dutchman
1965	M. McNamara	M. Rimmer	Albacore
1966	B. Ellis	K. Ellis	Merlin Rocket
1967	W. P. Bacon	M. McNamara	Enterprise
1968	R. Hennessy	R. Michael	Enterprise
1969	M. B. Rimmer	R. J. Suggitt	Enterprise
1970	R. J. Suggitt	Miss M. Watcham	Heron
1971	F. Williams	R. Sheffer	Merlin Rocket
1972	P. Milanes	P. Nash	Flying Dutchman
1973	N. Martin	Caroline Lougher	Nat. Firefly
1974	P. Crebbin	A. Landamore	Nat. Albacore

In 1974 the Royal Lymington Yacht Club started a 'Congressional Cup' series for IOR class leading helmsman by invitation. Match races were sailed using Contessa 32 yachts under an American tournament system.

Soling class (in the act of gybing)

Half Ton Cup boats designed to the IOR, usually about 30 ft, head out for an offshore race in the 1973 event

Half Ton Cup boats in the Baltic

CLASSES

All yachts and dinghies which race, and many which only cruise, belong to a 'class'. But it is a word without a single technical meaning: the variations of it become clear as it is used to refer to various boats. The boat which did not belong to any class at all would be said to be a 'one-off' and would not participate in racing. There may of course be another boat exactly like it, a builder having turned out successive boats one after the other and making them the same, but even in this case, it is seldom that such a group is not given a class name.

The types of classes can be found under one of these headings:

One-design class
Restricted class
Formula class
Handicap class
Builder's or commercial class

In each of these classifications can be found International classes – for instance these might be one-design or formula – National classes and then numerous unofficial classes.

To define the different types of class: the 'one-design' is strictly controlled by class rules (governed by an owners' association) which ensure that the hull is exactly the same shape, the weight distribution in it is also uniform and the same goes for the sail area and rig generally. One-design classes vary in their strictness: one class may insist that sails are all bought from the same sailmaker and then dealt out after a lottery, others may merely – and this is more usual – lay down measurement instructions and then the sails are inspected and stamped by the official class measurer. Uniformity would be ensured by regulations on length of luff, length of other edges, size of headboard, thickness of cloth and length of battens. Some classes allow tolerance in such things as deck fittings, others may not. In theory, a one-design class should be the cheapest racing class to sail, because changes cannot be made at great expense to outclass other boats in the class. The keenest owners will, however, attempt to develop those aspects of the boat which the class rule leave free: these could include the shape and number of sails, bottom surface, cross-section of rudder blade (all depending on the particular rules of the class).

'Restricted classes' have traditionally been more popular in Britain than elsewhere (and the USA was the breeding ground of the one-design, though one-designs are now known world-wide). In a restricted class, rules are drawn up to limit the length, weight, sail area and perhaps certain parts such as the rudder with its dimensions, mast height, height perhaps of spinnaker halyard block on the mast, and there may be rules about expensive types of gear such as winches and electronic instruments. But within this framework, the boat can be designed and craft of the class will, therefore, differ in hull shape, proportions of rig and materials used. Over the years the appearance of the latest boats in the class will therefore change. An example of a restricted class is the International Fourteen Foot Dinghy class.

A 'formula class' goes further along the path of design freedom in having a rating formula within which the designer can produce the fastest yacht he can. But after building, the yacht must be measured and the parts of the formula are combined to give the required rating. This will be a maximum and if the result is over the rating, the yacht is not 'in class'. The rating will usually be given in a linear measurement such as specifically in feet or metres, for example, the 12-metre. Details of the evolution of rating rules are given in Section 3. The yachts of the class may well look quite different due to the different interpretation of the rating rule by designers. This way of equating boats for racing is the most expensive owing to the risk of a new design quite outdating existing ones. This happened particularly earlier in the century when the control of rating rules was crude. There may be additional restrictions as in a restricted class on such things as minimum or maximum beam and height of mast and most likely control of 'exotic materials' (for example using uranium in a ballast keel to get weight very low). The Olympic Games featured almost all formula classes in the early days, but the last formula class in the Olympics was the 5·5-metre which was withdrawn after 1968. New formula classes are now unlikely to be introduced, 5·5-metres and 6-metres being sailed in some few parts of the world by a few enthusiasts only.

The most widely used formula class today is that of the International Offshore Rule, which is liberal and allows boats of a wide range of type and size: further details of the formula are given in Section 5. However, the rule instead of producing boats which after measurement must fit a formula, measures a boat which then has a rating in feet or metres. Boats in an IOR race will be of varying ratings and require a time allowance (see later in this Section) to equate their results. In certain instances the IOR is used to measure boats to a rating which must not be exceeded and these are the Ton Cup class yachts. They have certain extra restrictions on accommodation and equipment, but their main characteristic is to comply with the rating and measurement of the International Offshore Rule within a certain rating (see p. 199).

A 'handicap class' is not really a class at all, but a grouping at a regatta or event. So a yacht would be in a handicap class at the regatta of, say, the Royal Cornwall Yacht Club, but when she left Falmouth to sail to her home port elsewhere she is not in a 'class'. In a handicap class, the yachts or dinghies are given time against each other, which is adjusted to find the winner after each yacht has crossed the line. They may be handicapped by some single form of measurement, perhaps made by the owner himself, or they may be handicapped by the club measurer on performance or reputation. Handicapping systems are discussed further below.

These yachts and any of the other categories may also be in a 'builder's or commercial class'. Here it is simply a nomenclature when a builder produces a series of yachts to a certain design. He calls them the 'such and such' class and this is a convenient name. Most of them are likely to be closely standard, but manufacturing modifications may change details or even major features from time to time – indeed this is the rule rather than the exception. So the class is not controlled and measured in the same sense as a one-design or even a formula or restricted class. The builder *may* be producing a class which is also a one-design. In this case he would check the measurements of each boat in accordance with class rules when it left the yard. (It would also be checked by an official measurer before being allowed to race in the class.)

'International classes' are, strictly speaking, those that are officially recognized by the International Yacht Racing Union. The correct name of such classes contains the term 'International', for instance 'International

Dragon'. There are numerous other classes, however, that are international in the sense that the boats of the class are sailed and raced in many different countries and even recognized by the national authorities in particular countries. There may also be co-ordination of rules, measurement and so on. The Wayfarer class for instance is organized and sailed in Britain and widely in North America, and the associations in both parts of the world co-ordinate their handling of the class. But it is not an International class of the IYRU. The Cherub is another class in the same situation, having originated in New Zealand, but spread to be popular in Britain. In 1969, the Cherub class applied to the IYRU for international status, but this was turned down because such an application has to be supported by at least four national authorities and this did not occur.

International classes

Only 25 classes are internationally recognized today by the IYRU. They are listed below. The IYRU grades its recognized classes into A and B. Group A is for 'modern high performance classes, incorporating the latest designs, features and techniques'. Group A includes all the Olympic classes. In the event of extra classes being allowed or changes considered, new Olympic classes would come from Group A. Group B is for 'classes designed more than 20 years ago which are not considered suitable for Group A; classes which have been granted international status because of popular support, or because they provide a good training for yacht racing'. In Britain the Royal Yachting Association recognizes 16 classes as 'National' ones. These have to comply with a certain standard and strictness in class rules and command sufficient popular support in the country.

INTERNATIONAL CLASSES OF THE IYRU

(The word 'International' is in front of each name for the correct title)

Class	Type (OD or restricted)	Length
C class catamaran	Catamaran restricted	
Cadet	Centreboarder OD	10 ft 6 in
Contender	cb OD	16 ft 0 in
Dragon	keel boat OD	27 ft
Enterprise	cb OD	13 ft 3 in
Finn	cb OD	14 ft 9 in
5-0-5	cb OD	16 ft 6 in
5·5-metre	keel boat formula	31 ft (approx.)
Fireball	cb OD	16 ft 2 in
Flying Dutchman	cb OD	19 ft 11 in
Flying Junior	cb OD	13 ft 3 in
470	cb OD	15 ft 5 in
420	cb OD	13 ft 9 in
14 ft dinghy	cb restricted	14 ft 0 in
Lightning	cb OD	19 ft 0 in
Moth	cb restricted	11 ft 0 in
Optimist	cb OD	7 ft 7 in
6-metre	keel boat formula	33 ft (approx.)
Snipe	cb OD	15 ft 6 in
Soling	keel boat OD	26 ft 6 in
Star	keel boat OD	22 ft 8 in
Tempest	keel boat OD	22 ft 0 in
10 square metre/	cb development	
Tornado	catamaran OD	20 ft 0 in
Vaurien	cb OD	13 ft 3 in

National classes

The classes have the word 'National' as part of their name, for instance 'National Scorpion'. The RYA also has 90 class associations affiliated to it, apart from the National class associations and British associations of the International classes. These affiliated classes are owners' organizations, which have the general support of the national authority, but no special technical qualifications for recognition. Some are not racing boats, some are centreboarders, others are cruising boats whose organization may have started as a commercial class.

The national classes are the following:

Class	Type (OD or restricted)	Length
National Albacore	centreboard OD	15 ft 0 in
National Eighteen foot	cb OD restricted	18 ft 0 in
National Firefly	cb OD	12 ft 0 in
National Flying fifteen	keel boat OD	20 ft 0 in
National Graduate	cb OD	12 ft 6 in
RYA National Hornet	cb OD	16 ft 0 in
National Merlin Rocket	cb restricted	14 ft 0 in
National Osprey	cb OD	17 ft 7 in
National Redwing	cb OD	14 ft 0 in
National Scorpion	cb OD	14 ft 0 in
National Shearwater catamaran	catamaran OD	16 ft 6 in
National Solo	cb OD	12 ft 5 in
National Squib	keel boat OD	19 ft 0 in
National Swallow	keel boat OD	23 ft 6 in
National Swordfish	cb OD	15 ft 6 in
National Twelve Development Association	cb restricted	12 ft 0 in

In the USA there are no 'nationalized' or 'recognized' classes. It is up to clubs and associations to adopt whatever class they wish. Some classes have the word 'national' in front of them and this implies they are widespread, but confers no special status.

Numbers

The number of classes of racing boat in the world (international, local, commercial) is about 950. But new ones are being created, class associations forming and old classes fading at any particular moment. It is not possible to divide them by countries, because of immense overlap, even outside the official international classes. However, some 164 classes still active have originated in Britain and 525 in the USA; the balance originated elsewhere.

The number of centreboard sailing-dinghies in the world is in the region of 1 000 000 (1975).

As anyone can buy plans, or even borrow them and then build a particular dinghy without any further registration or notification, no one can say what the numbers really are. Communication is better in some classes than others; builders of one-designs may go into liquidation, records of early boats get lost; some boats go on sailing while others may be destroyed or left to rot, while the boat so lost remains registered. In most countries, small sailing boats are not docketed like houses or motor vehicles. About specific classes, however, there is significant and useful information.

Sunfish class.

Right
Sunfish, numerically largest class of sailing boat in the world.

The numerically largest class of sailing boat in the world is the surfboard-type single-sail centreboarder, the Sunfish. Its total number* in September 1974 was 140 000. Of these 130 000 are in the USA and the remaining 10 000 in other parts of the world. The class association claimed that 40 000 were active in 1974. In 1957 the design was switched from wood kits to glass-fibre production and the number of early home-made wooden ones is not possible to calculate accurately.

The Sunfish is of foam-filled GRP, length 13 ft 10 in, beam 4 ft $0\frac{1}{2}$ in, sail area (single lug) 75 ft², hull weight 139 lb. It can support a crew weight of 500 lb.

The first sailing surfboard was invented by Alex Bryan and Cortland Heyniger at Waterbury, Connecticut, and the first commercial class begun was the Sailfish. The Sunfish was introduced in 1951 and was first sold in kit form. When GRP boats were readily available in 1958 the class entered its highest growth rate. There has been a USA National Championship since 1961. The class was the vehicle for the world's most successful helmsman in a single class.

The numerically largest two-man centreboard dinghy class is the Mirror. On 16 September 1974 it had 47 589 boats registered: about 41 000 of these are in Britain. In the early 1970s, the class was increasing at a rate of 6000 per year and passed the 50 000 mark in June 1975. Though it is not recognized by the IYRU, there is a Mirror International Association with 13 national associations. The Mirror was designed by Jack Holt and Barry Bucknall and launched by the mass-circulation London *Daily Mirror* in 1963. It is of a special build-her-yourself design using plywood, glass fibre and stitching, the carrying capacity is good and there always have to be red sails – LOA 10 ft 10 in, beam 4 ft $7\frac{1}{2}$ in, weight 135 lb, sail area 69 ft².

* Current statistics unless otherwise stated are taken in this Section as at September 1974.

A single-handed OK class is the numerically largest single-handed centreboarded dinghy.

The numerically largest single-handed centreboard dinghy for adults is the OK class. There are 12 000 in the world, mostly in Scandinavia. There are 1790 in Britain. The boat is the most widespread single-handed class, being raced regularly in 36 countries. It was designed in 1957 by Knud Olsen of Denmark – LOA 13 ft 1½ in, beam 4 ft 8 in, weight 200 lb, area of single sail 90 ft².

The numerically largest keel boat class is the Star. There are 5000 in the world, 2000 of which are in the USA. This is an historic boat, the shape of the hull being in essence the same as when it was first produced – in 1911! They first sailed as a class of 22 in Long Island Sound in May of that year. The class association was formed at a meeting in New York in 1922. The sliding gunter rig was changed to Bermuda in 1921. In 1929 the boom was shortened and the mast lengthened, but the mainsail is very big and the jib small by today's standards. GRP hulls were authorized in 1967 and aluminium spars on 'bendy' rigs in 1971. The class was the first one-design to be in the Olympics and was in from 1932 to 1968.

The world championship has been held 35 times in North America, 10 times in Europe, 5 times in South America. Europeans have won the championship 14 times. The oldest winner was Charles de Carderas, 51, in 1954. The youngest was Malin Burnham, 17, in 1945. Two men each won three championships, Lowell North (USA) and Agostino Straulino (It.).

The first yachts to have flexible rigs were in the Star class. The Stars also had the first kicking strap (boom vang), circular vang tracks and were the first class to be fitted with self-bailing devices.

Class superlatives

The greatest class in value lies among boats of larger size because cost rises at greater than a linear scale owing to the complexity of bigger boats and the fact that displacement is a cubic measure. The 50 000 boats of the Mirror class are about £175 000 ($350 000) in value; the 13 000 boats of the International Offshore Rule class are the highest value of any single class – about £225 000 000 or $450 000 000.

The most expensive class of racing yacht is an International Offshore Rule boat of maximum rating, 70 ft. Probably two-masted and LOA about 78 to 80 ft with a crew of 15 to 18, the price (in 1975) ready for racing would

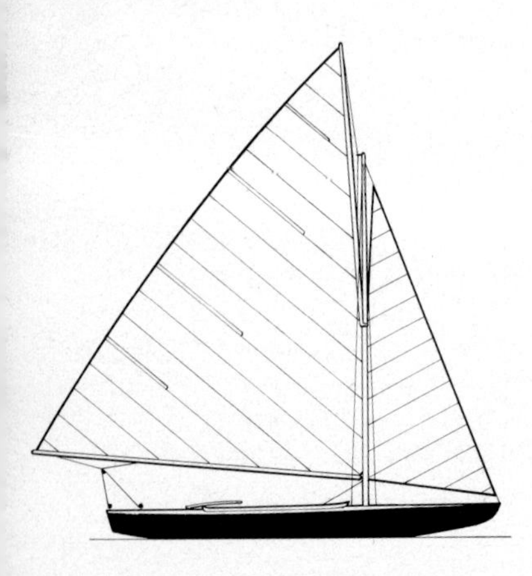

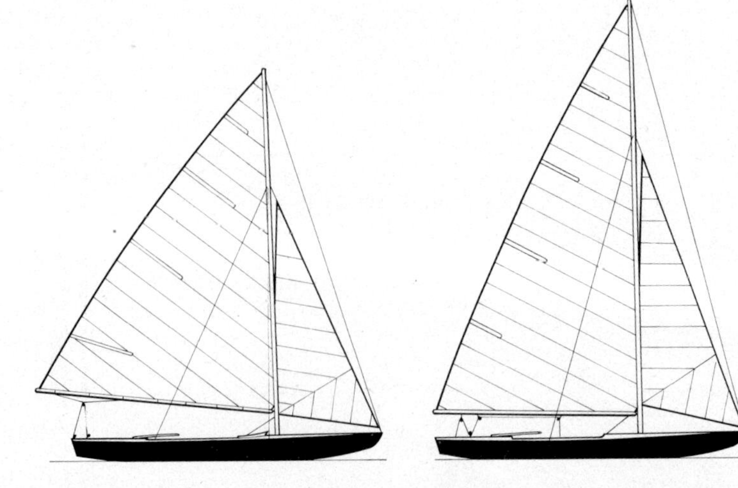

Below
Star class rigs. Left, Gunter rig of 1911 with long boom. Centre, 1921 rig with Bermuda mast first adopted but sails still of the previous shape. Right, 1929 rig which remains to the 1970s. This still has a very large mainsail by modern standards, but numerous control systems for shape and trimming have been introduced: metal spars were first allowed in 1971.

Star class yachts today. Note modern flexible rigs, but still the old proportions of mainsail and scow-like hulls of 60 years ago.

be about £200 000. The most expensive keel boat is the International 12-metre used for the America's Cup. The running costs of a season's campaign are difficult to separate from initial price. To build and campaign the winning defender *Courageous* in 1973 cost $1 500 000. *Intrepid*'s campaign (she was already built but was modified) cost $830 000.

The largest class yachts are those of the maximum size to the IOR or among keel boats, a 12-metre.

The smallest class boat is the International Optimist class which is 7 ft 7 in, beam 3 ft $8\frac{1}{2}$ in, weight 77 lb, area on single sail 39 ft^2. It is also the numerically largest centreboard sailing-dinghy: 105 000 are claimed by the class, of which 55 000 are registered. It is the cheapest racing class at under £100. The hull material is marine ply or can be GRP. The dinghy is designed for single-handed sailing by children, but is really too small for those over 15 years old and this is the maximum age for racing. It was designed by Clark Mills of Florida.

The fastest class boat regularly racing is the International Tornado catamaran one-design. It has the lowest Portsmouth Yardstick (63) – which implies highest speed – of any one-design class. It is the newest Olympic class, to appear for the first time at the 1976 Olympics. The world total (September 1974) is 1911 boats with 211 in Britain. Designed in 1966 by Rodney March (Brit.) of wood or GRP, the catamaran has LOA 20 ft 0 in, beam 10 ft, weight 276 lb, sail area 235 ft^2. A standard boat at speed trials in Portland under RYA supervision was timed at 20 knots.

The slowest class dinghy is not a claim that would ever be made! The shorter the slower, so the Optimist at 7 ft 7 in will be slow in absolute terms. It has the biggest Portsmouth Yardstick at 144. It has to be boxy so small wavelets soon reduce it to a crawl going to windward. Slowness for size is another matter and no doubt many a heavy cruising centreboarder which is undercanvassed is slower for its length.

X One-Designs racing. They are the oldest design of British keel boat still racing.

The world's oldest one-design class is the Water Wags of Dublin Bay founded in 1887. The class is still racing and has done so for every season except 1914–18. The annual subscription remained at ten shillings (50p) from 1887 until 1975 when it had to be raised.

The oldest one-design class still racing in the USA is the Inland Yachting Association A-Scow. First designed and built by John Johnson in 1897, the boat is a flat-hulled type with large sail area which takes a crew of five for sitting out. A class of up to 30 still race on the Great Lakes. With LOA 38 ft 6 in, beam 8 ft 6 in, draught 3 in (!), sail area 557 ft^2, the wood hull has foam buoyancy and weighs 1850 lb – price is about $10 000.

The first specifically organized one-design class in Britain was the Solent One Design which was created by the Solent Sailing Club in 1893 because 'This meeting views with the greatest concern the ever increasing expenditure attendant on small class racing on the Solent. It is strongly of the opinion that the sport should be conducted in a manner suitable to the means of the majority, and not merely with a view to the encouragement of the few who can afford year after year to build a new boat.' The boats were built in 1895–6, gaff cutters with bowsprit, LOA 33 ft 3 in. One-designs proliferated in the next years, with more than 20 starting in British yachting ports from 1895 to 1905.

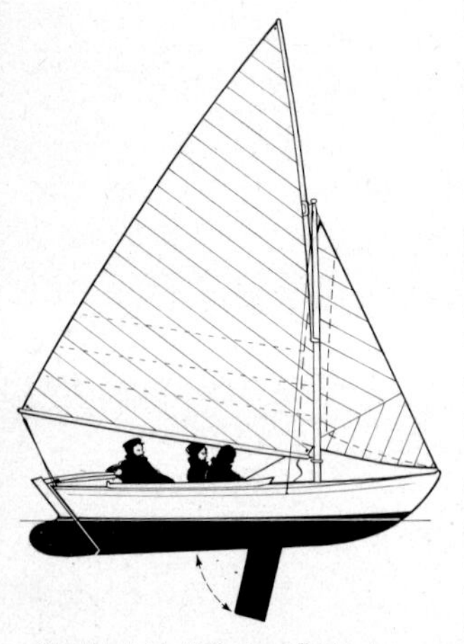

Seabird Half Rater. Oldest one-design class still racing in Britain.

The first US one-design class on salt water was the New York Yacht Club One Design which began in 1900. It consisted of four 70 ft yachts! This is the largest one-design class in hull size ever to have existed. It was designed by Nathanael Herreshoff and raced mainly off Newport, R.I., where the homes of the great millionaires were; in the class were Cornelius Vanderbilt and Harry Payne Whitney. In 1905, Herreshoff designed the New York Yacht Club 35 ft one-design which was to last for 40 years. The 'New

York 35' became a standard for judging other yachts. But on a national scale in the USA or Britain, one-designs were few, though many local one-design classes grew. Rated classes were used for national and international racing. The Star class was the first one-design to make an impact across the USA and then not until the 1920s.

The oldest one-design class in Britain still sailing and racing is the Seabird Half Rater. It was designed in 1898 by Herbert G. Baggs, the first were built in 1899 at a price of £34 17s 6d each – LOA 20 ft 0 in, LWL 16 ft 3 in, beam 6 ft 0 in, weight 2240 lb. Under the length and sail area rule in force in 1900, they rated at 0·5, but they have usually raced as a class.

Various clubs raced the boats in north-west England, but they are now sailed at Abersoch, North Wales and in the Mersey. Since 1899, 90 boats have been built and 69 still exist. The boats are still built in traditional materials and generally the original clinker construction has not been changed – but the cost has changed: to £2000, 57 times the price 76 years ago.

The oldest British keel boat class still racing regularly is the X One-Design. There are 162 still racing on the south coast of England with fleets at Poole, Yarmouth, Lymington, Cowes, Hamble, Itchenor – 174 have been built altogether so the wastage is small and many are very old.

The first boat was designed by Alfred Westmacott of the Isle of Wight and built in 1909 for less than £50. Up to 1914 10 yachts were built, up to 1939 a total of 35. The vast majority of this old design were, therefore, built after 1945. From 1928 Bermuda rig was optional instead of gaff, which it steadily superseded. The largest number in a single regatta has been 70 boats at Cowes Week 1973.

Ten X boats owned by the Kuwait Oil Company are at the Cumberland Yacht Club, Kuwait. Several X boats are sailed at Haifa, Israel.

Mistletoe, boat sail No. 1, built in 1909 is still sailing, owned by D. B. Payne and races in the Itchenor division of the class. No X boat is known to have capsized or sunk when sailing (unlike a number of other keel boat classes). Several of the class have made open sea passages: X67 was sailed single-handled from Falmouth to Bordeaux and Bordeaux to Hamble. The dimensions are LOA 20 ft 8 in, LWL 17 ft 6 in, beam 6 ft 0 in, draught 2 ft 9 in, sail area 184 ft². (In the USA 'an X boat' is 16 ft ILYA Scow designed in 1934. There are 450 of them.)

A pre-First World War class still raced is the Thames Estuary One Design, first built in 1912 to the design of Morgan Giles. The price of the first boat was £37 8s 2d (£37·41). Current GRP versions cost £900. The dimensions are LOA 18 ft, beam 6 ft, weight 800 lb. Thirty boats of the class race from the Alexandra Yacht Club, Southend-on-Sea, Essex.

A keel boat class still racing, yet with no hull having been built since 1938 is the Sunbeam class – LOA 26 ft 5 in, LWL 17 ft 6 in, beam 6 ft, draught 3 ft 9 in. The boat was designed by Alfred Westmacott in 1922 with Bermuda rig (since modernized). Thirty-nine boats were built between 1923 and 1938 and all but one still sail. There are 19 at Itchenor and 19 at Falmouth. The Sunbeam class has the record for participating in Cowes Week. From 1923 until today (except for 1940–5) the class has featured every year at Cowes and shows no sign of dropping out.

Bullseye. Designed by Nathanael Herreshoff in 1914 is the oldest American keel boat class still racing. It originated in and is still based at Buzzards Bay.

The oldest American keel boat class still racing is the Bullseye, designed in 1914 by Nathanael Herreshoff. There are 800 now sailing with headquarters at Buzzards Bay, Mass. The dimensions are LOA 15 ft 9 in, beam 5 ft 10 in, draught 2 ft 5 in, displacement 1350 lb. The class is now built in GRP

Uffa Fox's international 14-footer *Avenger* in 1928 with her prize-winning flags.

with modern rig and metal spars, but 200 early wooden boats remain in the class. The class in 1949 was the first in the world to change over to GRP. It is considered highly seaworthy for its size and there are boats in other parts of the US east coast, Bermuda and the Caribbean.

A close reproduction of the original 1914 gaff-rigged known as the 'Doughdish' (the Bullseye was often known as this) is still produced (in foam sandwich) for yachtsmen who want a period piece which sails.

The first centreboard dinghy class to be regularly raced internationally was the International Fourteen Dinghy class. It is still active. It is a restricted class – therefore the design changes within the rules. Maximum length is 14 ft, beam is between 4 ft 8 in and 5 ft 6 in, minimum weight is 275 lb. The boats must be open: this distinguishes them from most other classes today. The Fourteen was established as the first national dinghy class in Britain in 1923. It grew from the desire to bring together as a class various popular 14 ft sailing boats such as the then 'West of England Conference Dinghy', the 'Norfolk Dinghy' and 'Dublin Bay Water Wag': the last had been established in 1887. It became international in 1927 and from then on the most sophisticated racing-type centreboarder developed.

The designs of Uffa Fox of Cowes dominated the class from then until 1939. Uffa Fox's greatest achievement was the introduction of the planing dinghy in the class. His Fourteen *Avenger* was the most famous of all racing dinghies: in 1928 out of 57 starts, she finished 1st 52 times, and was 2nd twice and three times 3rd. Uffa sailed her to Le Havre in rough conditions taking 27 hours and then back again in 37 hours. Since *Avenger*, thousands of planing dinghies of all types have been built and sailed all over the world.

The first formal international contest in dinghies took place in Fourteens at Oyster Bay, Long Island Sound in September 1933. The British brought over boats for the Americans to sail, and the third nation was Canada, which entered with 14 ft cat-rigged (i.e. no jib) boats. The British won in their sloop-rigged boats (luckily for them there were fresh winds). The North Americans thereafter accepted the International Fourteen rules combined with their similar craft (although the Canadians did not agree fully with the international rulings until 1959). At times there has been other transatlantic disagreement and so variations in the boats.

Uffa Fox in *Brynhild*. He was first to win the New York Yacht Club canoe trophy for Britain.

The working parts of an International 14-footer. Note that it is still a completely open dinghy.

Right
International 14 ft dinghy class today with moulded plywood construction, metal spars and trapeze.

For instance the American Owners' Association allowed trapezes before the British. The International Association for the class now covers Britain, Bermuda, Canada, the USA and these four nations have regular international team races.

The Fourteen remains the most 'aristocratic' of dinghy classes (in breeding and longevity) and would never aspire to be large. In Britain about a dozen are built annually; 1000 have been built since 1922 in Britain and 1000 in the USA since 1933, 500 have been built in Canada. But being a restricted class, old designs become outdated and the active racing fleet in the world is at any time about 450 boats.

The only man to win a first-class dinghy championship in the same class as many as 12 times is Stewart H. Morris. He won the principal British championship trophy in the class, the Prince of Wales Cup in 1932, 1933, 1935, 1936, 1947 to 1949, 1957, 1960 to 1962 and 1965!

The numerically greatest official International class in Britain is the Enterprise centreboard dinghy. Until the advent of the Mirror it was the largest, being overtaken in 1968, but is now second with 17 750 boats at the end of 1974. Also started by a London newspaper (now defunct), its dimensions are LOA 13 ft 3 in, beam 5 ft 3 in, sail area 113 ft². It is sailed at over 450 clubs in Britain.

The greatest number of racing dinghies to take part in their class championship is 230 Enterprises. This was at the British National Championships of the class in 1973. Since then the number at the national championship has been reduced to 200 maximum.

The fastest-growing class in the world is the Laser single-handed dinghy. The GRP 'modern'-looking hull, seemed to catch the imagination of dinghy sailors all over the world in the early 1970s. The dimensions are LOA 13 ft 10½ in, beam 4 ft 6 in, weight 125 lb, area of single Bermuda sail 76 ft². It was designed by Bruce Kirby (Canadian) in 1970 and production started in Canada in 1971. In September 1974 there were 18 000 boats throughout the world. Though most are in North America, there is not a disproportionate number there. Sixteen hundred are in Britain. IYRU international status was granted in 1974.

C
K
2911
K6836
420
K20302

Far left
Aquacat: mass-produced 12 ft 2 in catamaran.

Left
The Cadet class, specially for sailors under seventeen.

Below left to right
Two pictures of the 420 class: most numerous on the Continent.

Firefly class.

OK dinghy.

The only classes of any sort of sailing boat in 1963 to be numerically greater than 10 000 were:

Optimist	17 500	Vaurien	13 200
Snipe	14 475	Enterprise	10 100

The numerical largest classes (all of which are one-designs except the IOR yachts) today are:

Class	**Length**	**Numbers**	**Year of origin**	**Area with greatest preponderance**
Sunfish	13 ft 10 in	140 000	1951	USA
Sailfish	13 ft 7 in	60 000	1947	USA
Optimist	7 ft 7 in	55 000	1948	(Everywhere)
Mirror	10 ft 10 in	47 589	1963	Britain
420	13 ft 9 in	28 900	1960	Continent
Snipe	15 ft 6 in	21 465	1931	USA and Continent
Laser	13 ft 10½ in	18 000	1970	(Everywhere)
Enterprise	13 ft 3 in	17 750	1956	Britain
Vaurien	13 ft 3 in	17 000		Continent
Aquacat	12 ft 2 in	14 000	1961	USA
OK	13 ft 1½ in	13 790	1957	Scandinavia
IOR	(about 20–80 ft)	12 850	–	(Everywhere)
Lightning	19 ft 0 in	12 700	1938	USA
GP 14	14 ft 1 in	11 800	1948	Britain
470	15 ft 4 in	11 500	1962	Continent
Flying Junior	13 ft 3 in	10 666	1955	Continent

Merlin Rocket class.

The numerically largest classes in Britain, which do not qualify for the above list are:

Class	Length	Year of origin	Numbers in Britain	Total in world
Cadet	10 ft 6 in	1947	5000	7000
Albacore	15 ft 0 in	1946	2000	6500
Graduate	12 ft 6 in	1952	2350	2565
Solo	12 ft 5 in	1955	1900	2600
Firefly	12 ft 0 in	1946	2700	3650
Fireball	16 ft 2 in	1962	5000	9716
Heron	11 ft 3 in	1951	5800	9003
Wayfarer	15 ft 10 in	1957	2600	4426

The restricted classes have the following latest sail numbers issued for new boats, but because the design is developed, older boats are disposed of and the numbers remain limited. The class, therefore, remains in existence, indeed the rules can be changed steadily to breathe in new life. A numerically small one-design class, by contrast, can die out because the design is no longer popular. Numerically big classes tend to continue through sheer weight of numbers, which is due anyway to widely based popularity. The restricted class latest numbers are:

Class	In Britain	Year of origin
International Fourteen	1040	1923
National Merlin Rocket	2970	1946
National Twelve	2808	1935

The numbers competing in class championship meeting are an indication of the activity and racing enthusiasm in the class. The Enterprise has the record in Britain, and probably in the world, at 230 starters in 1973. These figures apply to the championship meeting of a single class.

Increasing		**Mirror**		**5–0–5**	
1968	1972	80	116	70	120
1970	1974	127	200	80	116
Steady		**Fireball**		**GP 14**	
1968	1972	174	178	130	130
1970	1974	150	175	182	163
Erratic		**National Twelve**		**Merlin Rocket**	
1964	1972	200	162	135	143
1968	1974	101	128	225	138
1970		146			

Fireball class.

Single races with a mixture of classes have higher numbers starting, such as the Ensanada race in the USA, and in Britain the Round the Island (see p. 56) and Nore races which includes centreboard dinghy classes and all shapes and sizes of boat. National championships (in this case in Britain), surprisingly at first, have greater numbers than the corresponding international meeting. This is partly due to logistics and expense, some numbers per nation are limited by the class itself; sometimes Government and other grants to travel long distances with a dinghy are confined to a few people; some classes do not aspire to world championships, but have national meetings in countries where the particular class is well supported.

The biggest meetings by class have been:

Class	Number competing	Year
Enterprise	230	1973
Merlin Rocket	225	1970
Firefly	210	1963
National Twelve	200	1964
Mirror	200	1974
Fireball	194	1973
GP 14	185	1971
Scorpion	158	1974
OK	130	1970
5-0-5	120	1972

The biggest championship meeting in the USA in a single class was 203 in the Sunfish meeting in 1973.

Increasing numbers at class championship meetings are typically the Mirror (1968, 80; 1970, 127; 1972, 116; 1974, 200) and 5-0-5 (same sequence of years: 70, 80, 120, 116). Steady numbers are shown by the Fireball (174, 150, 178, 175); erratic numbers by GP 14 (130, 182, 130, 163) and Merlin Rocket (135, 225, 143, 138).

The centreboard class that developed fastest and most radically in design terms is the International Moth restricted class. One of the few restrictions is the length of 11 ft. The Portsmouth Yardstick has dropped from 109 in 1970 to 100 today, thus demonstrating its considerable actual increase in speed as a class against other racing dinghies. One of the reasons for this is that there is no weight restriction, so increasing use may be made of modern materials. The boats have developed large 'wings' for sitting out: these are banned in most other classes. Like all restricted classes there is considerable turnover, but there are about 3500 in the world.

The one-design class greatest in length is the International One Design. This is the undescriptive name of a wooden keel boat which is unusually elegant and was designed to be like the 6-metre class as it was developed in the 1930s. The dimensions are LOA 33 ft 5 in, beam 6 ft 9 in, draught 5 ft 4 in, displacement 7120 lb. There are about 150 in the world and they were designed in 1935 by Bjarne Aas (Norway). The boats had to be built in Norway and so have become increasingly old, though a number were built in the 1950s. There are fleets in Long Island Sound (where the standard has always been very high and has included winning America's Cup men like Arthur Knapp, Bus Mosbacher, Bob Bavier, and Ted Hood), Marblehead, San Francisco, Bermuda, Norway and until 1972 at Cowes, England. The boats from the last fleet have now been sold to owners based at Granton, Scotland. The world championship is run by

Flying Dutchman class

Finn single-handed Olympic dinghy class

470 class

International Moth with its characteristic 'wings'.

Right
The International One Design, greatest in length today.

bringing the helmsmen to where there is a fleet of boats and drawing for the use of a boat. Unlike centreboarders, it is impracticable to move the boats themselves.

The fastest-growing keel boat one-design in the USA is the Etchells 22. It was started in 1966 as a result of trials to choose a new IYRU three-man keel boat, in which trials it was unsuccessful, but it was promoted independently. The hull is GRP with LOA 30 ft 6 in, beam 7 ft 0 in, draught 4 ft 6 in, displacement 3325 lb. There are now 180, most of which are in the USA.

The only class in which three brothers are reported to have finished 2nd, 3rd and 4th in a world championship is the National One Design (USA). The brothers Paul (17 years), Mark (19) and Marty Makielskis (20) finished in that order and they come from South Bend, Indiana. There are 875 of these boats (800 in the USA): other boats are at Plymouth, England and in Northern Ireland.

In Plymouth the boats were first built in 1940 to an old set of plans and the first owners appeared to have no knowledge of the existence of the US class. It was not until 1960 that the orphans found that they had hundreds of cousins in North America and the class associations linked up. This is not as uncommon as it seems, though usually 'stray' class boats are cruising and racing in places such as Africa and the Far East without knowledge of the central association of the class. In England these boats are known as the 'Dolphin class' of the Royal Plymouth Corinthian Yacht Club and not

A mixed bag for launching: Moth, Firefly, 12 ft National.

'National' for obvious reasons. This centreboarder was designed by Bill Crosby in 1938 and is LOA 17 ft 0 in, LWL 10 ft 6 in (therefore it is unusual for a dinghy in having appreciable overhangs), beam 5 ft 8 in, minimum weight 400 lb.

The fastest 12 ft class is the Cherub with a Portsmouth Yardstick as low as 96. The class is the only New Zealand one to be adopted nationally in Britain. It is remarkably fast when reaching. It was designed in 1953 by John Spencer of Auckland. Dimensions are LOA 12 ft, beam 5 ft, weight minimum 110 lb. There are 2500, 450 of which are in Britain.

Another 12 ft class is the SigneT which has in a 24-hour race completed 102 miles at the Wyboston Sailing Club in June 1974. In 1963 four SigneTs crossed the Channel from Dover to Calais by way of a demonstration. The class was originally sponsored by the *Sunday Times* (hence the name). There are 697 boats in the world. The dimensions are LOA 12 ft 5 in, beam 4 ft 9 in, weight 190 lb.

The first two-man centreboarder designed specifically to be a high-performance racing boat for children was the Cadet. It was designed by Jack Holt and sponsored by *Yachting World* in 1947. There are 7000 boats in the world: about 2000 are in Britain. The class is particularly popular outside the traditional Western 'yachting' countries and behind the Iron Curtain. There are fleets not only in the USA, Canada, the Netherlands, Belgium and Italy but also in Spain, Portugal, Poland, East Germany,

Etchells 22 class.

Right
Flying Fifteen: British national keel boat designed by Uffa Fox.

Cherub class: the fastest 12 ft class which originated in New Zealand.

Yugoslavia, Czechoslovakia, South Africa and Hungary. The world championship was held in Poland in 1968. There are known to be a number of Cadets in the USSR, but the Russians have so far refused to join the International Association (though the USSR is a member of the IYRU). The dimensions are LOA 10 ft $6\frac{3}{4}$ in, beam 4 ft 2 in, weight 120 lb, sail area 55 ft^2. For four years running from 1953 to 1956 Ralph and Brian Ellis won the national championship in *Dial* (Cadet No. 999).

The Hornet is the only British national class to have nearly half the boats in the world in a Communist country – Poland. Designed by Jack Holt for *Yachting World* in 1951, it is the only class to allow either a sliding seat *or* a trapeze and it had a sliding seat from its inception, unusual in dinghy classes (this type of sliding seat is for sitting out athwartships to hold up the boat in a breeze). The dimensions are LOA 16 ft, beam 4 ft 7 in, weight 300 lb, sail area 121 ft^2. Of the 1950 boats in the world, 700 are in Britain and 800 are registered in Poland.

The most numerically popular keel boat class in Britain is the National Flying Fifteen of which there are 1200 in the British Isles and about 550 in the rest of the world. A strong fleet is in Northern Ireland. Designed by Uffa Fox in 1947, the boat was based in conception of the old small raters seen in the Solent at the turn of the century. The dimensions are LOA 20 ft,

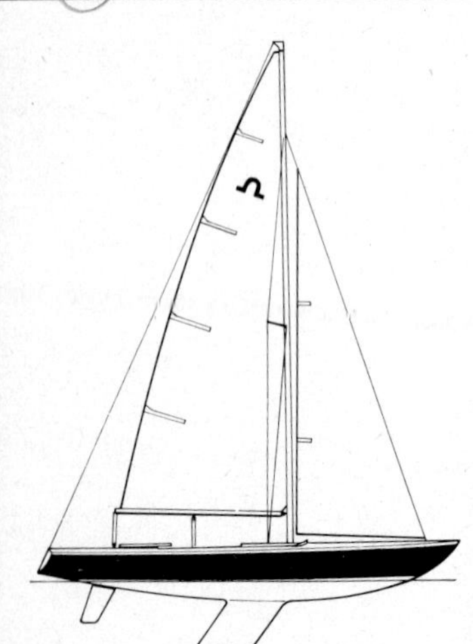

LWL 15 ft (hence the name: today it would be unlikely that a class would be called by any dimension other than its greatest, for example 20!), beam 5 ft, weight 675 lb. The class is sailed at 52 clubs in Britain and Ireland and has grown particularly rapidly at the beginning of its third decade. In 1972–3, 134 new boats were registered and in 1973–4 there were 112. The popularity is said to be in some measure due to being the only planing keel boat which it is not necessary to sit out or trapeze. It is, therefore, very suitable for older people; such people can more easily afford to buy one.

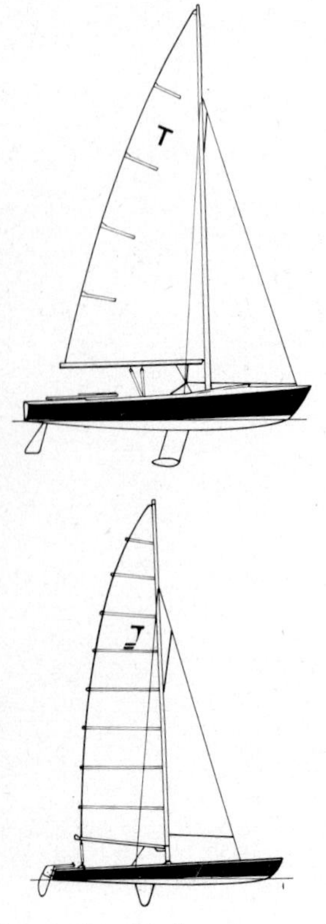

International Soling class, 26 ft 9 in. Three-man Olympic keel boat.

International Tempest class, 21 ft 11¾ in. Olympic two-man keel boat.

International Tornado class, 20 ft 0 in, Olympic two-man catamaran class. First appearance in Olympics 1976.

Olympic classes

The most specialized classes are those used for the Olympic Games. This is because of development rather than basic characteristics, although they have been chosen because they are high-performance boats that will be demanding in competition. They are not necessarily popular in terms of numbers, indeed clubs and individuals often avoid them because they fear that the 'hot shots' in the class will force development innovations that will too quickly outdate boats at club level. On the other hand the classes are widespread in the world as a whole because they are acknowledged for international competition and in many countries such classes are the only ones to receive Government backing and financial aid. The East Germans for instance have done very well in the Flying Dutchman class for a number of years from the mid 1960s and the Russians have often won one class or another in the Games themselves (in 1972 they won the Tempests). After many years of wrangling the International Yacht Racing Union has since the 1972 Olympics rationalized its classes to 'one three-man keel boat, one two-man keel boat, two two-man centreboarders, one single-handed centreboarder, one multihull class'. In the categories the classes are all one designs, the Soling, Tempest, Flying Dutchman, 470, Finn and Tornado.

'Olympic classes' with dimensions and specification; nature of origin and date; first date in Olympic Games; number in the world:

International 470 class, 15 ft 5 in. Olympic second two-man centreboard class, first appearance in the Olympics 1976.

International Finn class, 14 ft 9 in. Olympic single-hander.

Soling – LOA 26 ft 9 in, LWL 22 ft 2 in, beam 6 ft 3 in, draught 3 ft 3 in, weight 2277 lb, sail area 233 ft². Designed by Jan Linge (Norway). Strict glass-fibre mould for hull with metal spars. Keel fixed but it can be trailed behind a 2000 cc car with a suitable trailer. Chosen as a result of 'three-man keel boat' trials conducted by the IYRU in 1967 and selected from six other contenders and intended from thc beginning to be the successor in the Olympics to the older Dragon. First competed in the Olympics in 1972. There are 2256 in the world.

Tempest – LOA 21 ft 11¾ in, LWL 19 ft 3 in, beam 6 ft 2 in, draught 3 ft 7 in, weight 975 lb, sail area 247 ft². Designed by Ian Proctor (Brit.). Strict glass-fibre hull for mould with metal spars and special 'bendy' rig. The keel is retractable for trailing but fixed in one position for sailing. Chosen as a result of 'two-man keel boat' trials conducted by the IYRU in 1965, when she beat ten other contenders. After 11 years of controversy it replaced the Star class in the Olympics for 1976; both classes were in the 1972 Games. There are 658 in the world.

Flying Dutchman – LOA 19 ft 10 in, beam 5 ft 6 in, weight 435 lb, sail area 200 ft². Designed by Uffa van Essen (Neth.) in 1951, it is a one-design which allows latitude in fitting out. It sailed as a class before competing in IYRU trials and being selected as a high-performance two-man centreboarder. First used in the 1960 Olympic Games. Built of wood by hand by specialist builders. There are about 3700 in the world.

One of the few racing dinghies to be preserved historically is the British Flying Dutchman *Superdocious* in the National Maritime Museum, Greenwich. In the hands of Rodney Pattisson, MBE and Iain Macdonald-Smith, MBE, the boat was winner of the Gold Medal in the class at the Olympic Games, Acapulco in 1968; British champion in 1968, 1969, 1971; European champion 1968, 1971 and world champion 1969, 1970 and 1971.

470 – LOA 15 ft 5 in, LWL 14 ft 5½ in, beam 5 ft 7 in, weight 260 lb, sail area 143 ft². A strictly controlled one-design of glass fibre and light weight for top performance. Designed by Andre Cornu (Fr.), the class was well

Left above
Latest two-man dinghy to join the Olympic classes: the 470.

Right
Flying Dutchman. The demanding 19 ft 10 in two-man Olympic dinghy.

established before being adopted as an Olympic class, after debate as to whether this second two-man centreboarder, a French one, should be chosen, or whether it should be the British Fireball. The first Games in which it participates will be 1976. There are about 22 789 already sailing in the world.

Finn – LOA 14 ft 9 in, LWL 14 ft $2\frac{1}{2}$ in, beam 4 ft $11\frac{1}{2}$ in, weight 145 lb, area of single sail 108 ft^2. Designed by Rickard Sarby (Fin.). The boat is produced by various builders in different parts of the world to strict one-design dimensions, though as with all class rules this has caused difficulties in interpretation over the years. It was introduced by Finland after trials in Scandinavia for the Olympic Games in Helsinki in 1952, because there was no existing suitable single-handed class for the Games. It is a taxing boat to sail and it is usually reserved to those with Olympic aspirations, thus there are only 400 in Britain. The number of boats is unknown to the class authorities. The USSR, which will not reveal its Finn fleet, may have over 3000.

Tornado – As the fastest one-design class, the particulars have been given on p. 81. It was designed in 1966 by Rodney March (Brit.), and was a clear selection at 1967 trials held by the IYRU for an Olympic multihull. It was due to appear at the 1976 Olympics as the first multihull class at the games. There are 1911 boats in 31 countries.

OLYMPIC GAMES

The first yachting events in the Olympic Games took place in 1900 on the Seine at Meulan, Paris, since the main Olympic events were at Paris that year. The first Olympic Games had been in Greece in 1896, but had not included yacht racing. The classes at the 1900 Olympics were quite different to any subsequent ones, having an 'open class' using time allowance and then classes by limited tonnage (as rated at the time). The six classes by tonnage were $\frac{1}{2}$ ton, $\frac{1}{2}$ ton to 1 ton, 1 ton to 2 tons, 2 to 3 tons, 3 to 10 tons and there was a 10 to 20 ton class (using time allowance where required). This last class was sailed at the mouth of the Seine at Le Havre, because the river at Paris was clearly too narrow for such vessels. This last class was also sailed more than two months later than the others – the small classes being in late May 1900 and the big classes in early May. Because of the open class, the British yacht *Scotia* rated at half a ton was able to win a Gold Medal in that class and in the class of yachts measuring $\frac{1}{2}$ ton. This is the only time a yacht and its skipper have won two classes in an Olympic Games and can never be repeated, owing to modern classification.

Scotia was designed by Linton Hope and crewed by Lorne C. Currie, J. H. Gretton and the designer. The other class winners were France in the $\frac{1}{2}$ to 1 ton class, and 10 to 20, Britain in the 2 to 3 ton class, Germany in the 1 to 2 ton class and the USA in the 3 to 10 ton class. The only other nation to appear at all among the medallists was Switzerland which was second in the 1 to 2 ton class.

Yachting events featured in the Olympics at each games from 1908, when the main Games were in London and the yacht races at Cowes. In 1904 the Games were in the USA and there were no yacht races: at that date there were no international classes and there existed particularly wide variations between American and European racing boats. By 1908 there was in existence the first International Rule of rating (see Section 5) and the classes were based on this; they were the 12-metre, 8-metre, 7-metre and 6-metre. The table, page 101, shows classes then and subsequently in

Early Olympic Games sailed in formula metre classes. 8-metres here were of the type used in the 1908 Olympic Games.

yachting events in the Olympic Games. This shows that the two classes used at the greatest successive number of Games were the Star and the 6-metre (although the rating rule of the 6-metre was changed and therefore the design of the yacht, though the concept remained the same).

This table also shows that the heavy day keel boats have given way to light two-man keel boats and centreboard boats; that 1920 had the largest number of different classes, but not the greatest number of yachts; that 1976 will contain the fastest boat to have competed in the Olympics, the Tornado catamaran; 6 classes have taken part in one Games only (not including 1900). The single-handed dinghy class has changed, though from 1952 it has become stabilized in the Finn one-design. In 1920 the 12 ft International class was used, in 1936 the Olympic monotype and in 1948 the National (British) Firefly.

The largest yacht ever used in the Olympic Games was the 12-metre in 1908, 1912 and 1920.

The largest number of yachts to compete in any one class was 36 in the Finn class in 1968.

The largest number of boats to compete in an Olympic regatta was 152, the total number in the Olympics of 1972. This was 35 Finns, 29 Flying Dutchman, 26 Solings, 23 Dragons, 21 Tempests and 18 Stars.

The yachtsman to win the greatest number of gold medals is Paul Elvström of Denmark. He won the single-handed class in 1948 (in Fireflies) and in

Paul Elvström

1952, 1956, 1960 (in Finns). At the time this record of four successive gold medals was unsurpassed in any sport and in yachting it remains unapproached. Elvström came 4th in the 1968 games in the Star class (20 boats competing).

The largest number of countries to have competed in an Olympic regatta was 46 in 1960. This also had the greatest number of competitors, 288 – all male. Women have been in crews of Olympic boats, but the greatest number in a single year was 3: this was in both 1936 and 1952. The introduction of even lighter and more demanding classes in recent years means that this figure is unlikely to be exceeded.

The country to have taken part in most Games is France, which has had one or more boats in every yachting Olympics, next come Britain and Sweden who have each omitted one year, 1912 and 1900 respectively. The USA competed in 1900, but not again until 1928 when it has entered a team on every successive Games.

By far the most elaborate yacht harbour constructed for the Games was that in Kiel Fiord, Germany, for the 1972 events. The harbour remains after the games for regattas and yachting facilities generally. As built, the main block ran for 500 yards parallel to the waterfront and contained the central administration, race committee offices, jury-room, boat measurement hall where sails and hulls were checked, restaurants, swimming-pool and communications centre. 'Olympic village' accommodation was provided in 32 bungalows and 2 apartment blocks for competitors including flats for each yacht crew and offices for team managers.

Accommodation for the Press was on the top three floors of the main building, each cabin being fitted with typewriter and television set. From the balcony windows the Press could see all the race-course at sea, but 26 Press boats were available to take reporters and photographers to follow each event. Fourteen other vessels were available to take up to 4000 general spectators. All races were shown on local television and each boat photographed rounding each mark as evidence in the event of protests. Because of the nature of the Olympics and public interest the shore base at Kiel had more different passes and permits than any other organized yacht regatta. In the Olympics as a whole there was reaction to the size and elaboration of the organization of the 1972 Games at Munich (and at Kiel) and it is possible that future games will be simpler – but the intensity of yachting competition and developing techniques mean the scope for reduction in organization is limited.

The country to have won the most gold medals in yachting up to 1972 was Norway with 15; then followed the USA 12, Britain 12, Sweden 8, France 6. Only three of the Norwegian medals were gained from 1948 onwards and the medals came from the Norwegian dominance in the heavy keel boats used prior to then. This also applies, but to a lesser extent, to Britain and France, leaving the USA with clearly the most medals since 1948.

The greatest number of gold medals to be won by any one country in a single Olympic regatta since 1948 is 2. It has happened in five of the seven occasions. The USA has won 2 in 1948, 1952 and 1968. Sweden won 2 in 1956; Australia won 2 in 1972.

Since 1948 the total score of gold medals has been the USA 9, Denmark 5, Australia, Britain, Norway, Sweden, the USSR 3, Bahamas, France, Greece, New Zealand, Italy 1 each.

CLASSES OF YACHT AT SUCCESSIVE OLYMPIC GAMES

Year	Heavy fixed keel boats												Light two-man keel boats		Two-man dinghy	One-man dinghy	Light two-man dinghy	Multihull
1900	6 classes: 0·5 to 20 tons by rating and one open class																	
1908	12-m		8-m			7-m		6-m										
1912	12-m	10-m	8-m					6-m										
1920	12-m (2)	10-m (2)	8-m (2)	30	40	7-m	6·5-m	6-m								1		
1924			8-m					6-m								1		
1928			8-m					6-m								1		
1932			8-m					6-m					★			1		
1936			8-m					6-m					★			1		
1948								6-m		Sw		D	★			1		
1952								6-m	5·5-m			D	★			Fin		
1956									5·5-m			D	★		12S	Fin		
1960									5·5-m			D	★		FD	Fin		
1964									5·5-m			D	★		FD	Fin		
1968									5·5-m			D	★		FD	Fin		
1972											S	D	★	T	FD	Fin		
1976											S			T	FD	Fin	470	Torn

12-m, 8-m, etc.: 12-metre, etc. boats to International Rule
(2): Old and new rules at that time (two classes)
30, 40: 'square metre' classes
Sw: Swallow OD
D: International Dragon
S: Soling class
★: Star class

12S: 12 square metre sharpie class
FD: Flying Dutchman
Fin: Finn single-handed dinghy
470: 470 class
Torn: Tornado catamaran
1: Various single-handed dinghies

Olympic yachting has gradually reached a certain pattern of race organization. It consists of a race each day for seven days, but with a lay day or days in the middle. Each competitor is allowed to discard his worst performance and therefore counts the best six. Points are awarded, depending on placing in each race, to find the winner. The regatta is organized with one area for Solings, one for Flying Dutchmen and Tempests, one for Finns and one for Tornados. An Olympic course is then laid in the circle allotted to these classes. The 'Olympic course' is frequently mentioned in yacht racing as it is used in other regattas all over the world.

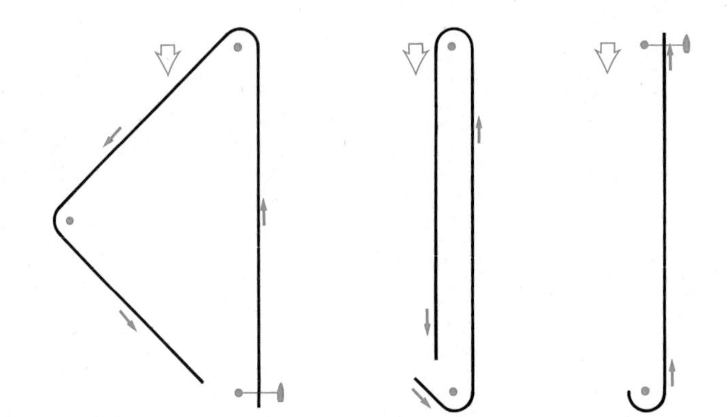

Diagram of the 'Olympic course' first leg is a beat to windward followed by two reaches; then a beat and a run; finally a beat to the finish. Marks are left to port. The course cannot be set until the wind direction is known.

After a number of years the Olympic course has stabilized to the following form (see diagram above) The circle for the class has a certain diameter. For the Solings this is 2·7 miles; for the Tempests and Flying Dutchmen the same. For the Finns it is 1·9 miles. The committee boat lays the start-line at the leeward point of the circle. The first leg is to windward to the opposite diameter; in other words the length of the windward leg is the diameter and varies with the size of yacht. The yachts round the windward mark, leaving it to port and all subsequent marks to port. The course is then to the reaching mark, then to the leeward mark which was originally the outer distance mark of the start. Then the fourth leg is dead to windward (repeating the first leg), the next leg is from the windward mark to the leeward mark (dead downwind if the wind has remained constant in direction). During this leg when all yachts have rounded the windward mark, the committee boat gets under way and keeping clear of the course, steams to the windward mark where she takes up position to starboard of it (as seen by a yacht approaching from leeward) to form a finishing-line. The yachts then round the leeward mark and sail the last windward leg (so first, fourth and sixth are the same) to this finish which has been set up.

Thus the Olympic course has considerable emphasis on windward ability. If the course diameter is 2·7 miles then the yachts have sailed 8·1 miles dead to windward, 2·7 miles dead to leeward and 3·6 miles on reaches. If the wind changes after the yachts have rounded the windward mark the first time, the race committee can shift the windward mark so that the next two windward legs will be truly to windward in the direction of the new wind. Special signals would be flown by officials' boats to warn of this. It need hardly be pointed out that if the wind is uncertain, the conduct of Olympic courses becomes difficult. For this reason it tends to be used, apart from the Olympic Games, in major championships and trials only. At a lower level the committee can more easily set a satisfactory course utilizing existing marks in a locality. Usually some windward work will be introduced by choosing suitable legs.

The Olympic Games occupy an anomalous position in yacht racing, because only a few classes, or even types of yacht, are represented. However, Olympic classes have a status of their own which attracts the best racing helmsmen and crews. On Olympic years the six classes have no world championships, though there are international meetings. Though to gain a medal in the Olympic Games is a mark of the highest yacht-racing ability and is understood internationally, the standard across the whole field of an Olympic race may be lower than in a world championship of the class. The reason for this is that in the Olympic race there is only one boat per nation. In a certain class, nation A may have a very high standard and its top five boats might be better than more than half the fleet in an Olympic race, which has entries like nation B. Now Nation B has only a few boats of this class in its country. They would only be club standard in nation A. In the same way national trials can be tougher in a country where the class is keenly sailed, than in the Olympic regatta itself.

The results of the Olympic Games were as follows. To begin with they are shown year by year, owing to the great variation in the classes used, but from 1924 the results are shown class by class.

Class	Number of countries	First three places	Skipper
		1900	
Open		(1) UK	Lorne C. Currie
		(2) Germany	Martin Wiesner
		(3) France	E. Michelet
½ ton		(1) France	Texier
		(2) France	Pierre Gervais
		(3) France	Henri Monnot
½–1 ton		(1) UK	Lorne C. Currie
		(2) France	Jacques Bandrier
		(3) France	E. Michelet
1–2 ton		(1) Germany	Martin Wiesner
		(2) Switzerland	Comte Hermann de Pourtales
		(3) France	Vilamitjana
2–3 ton		(1) UK	E. Shaw
		(2) France	Susse
		(3) France	Donny
3–10 ton		(1) USA	Howard Taylor
		(2) France	Maurice Gufflet
		(3) USA	Mac Henry
10–20 ton		(1) France	E. Billard
		(2) France	Duc Decazes
		(3) UK	Edward Hore
		1908	
6-metre	4	(1) UK	G. U. Laws
		(2) Belgium	Leon Huybrechts
		(3) France	Henri Arthus
7-metre	1	(1) UK	Charles Rivett-Carnac
8-metre	3	(1) UK	Blair Cochrane
		(2) Sweden	Carl Hellström
		(3) UK	R. Himloke
12-metre	1	(1) UK	Thomas Glen-Coats
		(2) UK	Charles MacIver

Class	Number of countries	First three places		Skipper
		1912		
6-metre	5	(1)	France	Amedee Thube
		(2)	Denmark	Hans Meulengracht-Madsen
		(3)	Sweden	Harald Sandberg
8-metre	4	(1)	Norway	Thoralf Glad
		(2)	Sweden	Bengt Heyman
		(3)	Finland	Bertil Tallberg
10-metre	3	(1)	Sweden	Carl Hellström Jr.
		(2)	Finland	Harry Wahl
		(3)	USSR	E. Belvselvsky
12-metre	3	(1)	Norway	Johan Anker
		(2)	Sweden	Nils Persson
		(3)	Finland	Ernst Krogius
		1920		
12 ft dinghy		(1)	Netherlands	Johannes Joseph Antonius Hin
18 ft dinghy		(1)	UK	F. A. Richards
6-metre		(1)	Norway	Andreas Brecke
6-metre (old rule)		(1)	Belgium	Emile Cornellie
6·5-metre		(1)	Netherlands	Johan Carp
7-metre		(1)	UK	Cyril Macey
8-metre		(1)	Norway	Magnus Konow
8-metre (old rule)		(1)	Norway	August Ringvold Sen.
10-metre (new rule)		(1)	Norway	Archer Arentz
10-metre (old rule)		(1)	Norway	Erik Herseth
12-metre (old rule)		(1)	Norway	Henrik Ostervold
12-metre (new rule)		(1)	Norway	Johan Friele
30 sq. metre		(1)	Sweden	Gösta Lundqvist
40 sq. metre		(1)	Sweden	Tore Holm

(In 1920 entries were very small with at least one sail-over)

RESULTS IN OLYMPIC CLASSES FROM 1924 ONWARDS.

	Year	Number of boats	First three		Skipper
8-metres	1924	5	(1)	Norway	August Peingvold Sen.
			(2)	UK	E. E. Jacob
			(3)	France	Louis Breguet
	1928	8	(1)	France	Donatien Bouché
			(2)	Holland	Lambertus Doedes
			(3)	Sweden	John Sandblom
	1932	2	(1)	USA	Owen Churchill
			(2)	Canada	Ronald Maitland
	1936	10	(1)	Italy	Giovanni Leone Reggio
			(2)	Norway	Olav Ditlev-Simonsen, Jr
			(3)	Germany	Hans Howaldt

	Year	Number of boats	First three		Skipper
6-metres	1924	9	(1)	Norway	Eugen Lunde
			(2)	Denmark	Wilhelm Vett
			(3)	Netherlands	Johan R. Carp
	1928	13	(1)	Norway	Kronprinz Olav
			(2)	Denmark	Niels Otto Möller
			(3)	USSR	Nikolay Wekschin
	1932	3	(1)	Sweden	Tore Holm
			(2)	USA	Frederick Conant
			(3)	Canada	Philip Rogers
	1936	12	(1)	UK	Charles Leaf
			(2)	Norway	Magnus Konow
			(3)	Sweden	Sven Salen
	1948	11	(1)	USA	Herman Whiton
			(2)	Argentina	Enrique Sieburger
			(3)	Sweden	Tore Holm
	1952	11	(1)	USA	Herman Whiton
			(2)	Norway	Finn Ferner
			(3)	Finland	Ernst Westerlund
5·5-metres	1952	16	(1)	USA	Dr Britton Chance
			(2)	Norway	Peder Lunde
			(3)	Sweden	Folke Wassen
	1956	10	(1)	Sweden	Lars Thörn
			(2)	UK	Robert Perry
			(3)	Australia	Alexander Sturrock
	1960	19	(1)	USA	George O'Day
			(2)	Denmark	William Berntsen
			(3)	Switzerland	Henri Copponex
	1964	15	(1)	Australia	William Northam
			(2)	Sweden	Lars Thörn
			(3)	USA	John McNamara
	1968	14	(1)	Sweden	Ulf Sundelin
			(2)	Switzerland	Louis Noverraz
			(3)	UK	Robin Aisher
Dragons	1948	12	(1)	Norway	Thor Thorvaldsen
			(2)	Sweden	Folke Bohlin
			(3)	Denmark	William Berntsen
	1952	17	(1)	Norway	Thor Thorvaldsen
			(2)	Sweden	Per Gedda
			(3)	Germany	Theodor Thomsen
	1956	16	(1)	Sweden	Folke Bohlin
			(2)	Denmark	Ole Berntsen
			(3)	UK	Graham Mann
	1960	27	(1)	Greece	Kronprinz Konstantin
			(2)	Argentina	Jorge Salas Chaves
			(3)	Italy	Antonio Cosentino
	1964	23	(1)	Denmark	Ole Berntsen
			(2)	Germany	Peter Ahrendt
			(3)	USA	Lowell North
	1968	23	(1)	USA	George Friedrichs
			(2)	Denmark	Aage Birch
			(3)	E. Germany	Paul Borowski
	1972	23	(1)	Australia	John Bruce Cuneo
			(2)	E. Germany	Paul Borowski
			(3)	USA	Donald Cohan

	Year	Number of boats	First three		Skipper
Stars	1932	7	(1)	USA	Gilbert Gray
			(2)	UK	Colin Ratsey
			(3)	Sweden	Gunnar Asther
	1936	12	(1)	Germany	Dr Peter Bischoff
			(2)	Sweden	Arvid Laurin
			(3)	Netherlands	Willem de Vries-Lentsch
	1948	17	(1)	USA	Hilary Smart
			(2)	Cuba	Dr Carlos De Cardenas
			(3)	Netherlands	Adriaan Lambertus
	1952	21	(1)	Italy	Agostino Straulino
			(2)	USA	John Reid
			(3)	Portugal	Joaquim De Mascarenhas Fiuza
	1956	12	(1)	USA	Herbert Williams
			(2)	Italy	Agostino Straulino
			(3)	Bahama	Durward Knowles
	1960	26	(1)	USSR	Timir Pinegin
			(2)	Portugal	José Quina
			(3)	USA	William Parks
	1964	17	(1)	Bahama	Durward Knowles
			(2)	USA	Richard Stearns
			(3)	Sweden	Pelle Pettersson
	1968	20	(1)	USA	Lowell North
			(2)	Norway	Peder Lunde Jr
			(3)	Italy	Franco Cavallo
	1972	18	(1)	Australia	David Forbes
			(2)	Sweden	Pelle Pettersson
			(3)	Germany	Willi Kuhweide
Swallow	1948	14	(1)	UK	Stewart Morris
			(2)	Portugal	Duarte De Almeida Bello
			(3)	USA	Lockwood Pirie
12 sq. metre Sharpie	1956	13	(1)	New Zealand	Peter Mander
			(2)	Australia	Roland Tasker
			(3)	UK	Jasper Blackall

Rodney Pattisson and Ian Macdonald-Smith, winners of the Flying Dutchman class in 1968. Pattisson won again in 1972.

	Year	Number of boats	First three	Skipper
Soling	1972	26	(1) USA	Harry Melges Jr
			(2) Sweden	Stig Wennerstroem
			(3) Canada	David Miller
Tempest	1972	21	(1) USSR	Valentin Mankin
			(2) UK	Alan Warren
			(3) USA	Glen Foster
Flying Dutchman	1960	31	(1) Norway	Peder Lunde Jr
			(2) Denmark	Hans Fogh
			(3) Germany	Rolf Mulka
	1964	21	(1) New Zealand	Helmer Pedersen
			(2) UK	Keith Musto
			(3) USA	Harry Melges Jr
	1968	30	(1) UK	Rodney Pattisson
			(2) Germany	Ullrich Libor
			(3) Brazil	Reinaldo Conrad
	1972	29	(1) UK	Rodney Pattisson
			(2) France	Yves Pajot
			(3) Germany	Ullrich Libor
Single-handed Dinghy class	1924	17	(1) Belgium	Leon Huybrechts
			(2) Norway	Henrik Robert
			(3) Finland	Hans Dittmar
	1928	20	(1) Sweden	Sven Thorell
			(2) Norway	Henrik Robert
			(3) Finland	Bertil Broman
	1932	11	(1) France	Jacques Lebrun
			(2) Holland	Adriaan Lambertus Maas
			(3) Spain	Santiago Amat Cansino
	1936	25	(1) Netherlands	Daniel Marinus J. Kagchelland
	Olympic monotype		(2) Germany	Werner Krogmann
			(3) UK	Peter M. Scott
	1948 Firefly	21	(1) Denmark	Paul Elvström
			(2) USA	Ralph Evans Jr
			(3) Netherlands	Jacobus Hermanus de Jong
	1952 Finn	28	(1) Denmark	Paul Elvström
			(2) UK	Charles Currey
			(3) Sweden	Rickard Sarby
	1956	20	(1) Denmark	Paul Elvström
			(2) Belgium	André Nelis
			(3) USA	John Marvin
	1960	35	(1) Denmark	Paul Elvström
			(2) USSR	Aleksandr Tschutschelov
			(3) Belgium	André Nelis
	1964	33	(1) Germany	Wilhelm Kuhweide
			(2) USA	Peter Barrett
			(3) Denmark	Henning Wind
	1968	36	(1) USSR	Valentin Mankin
			(2) Austria	Hubert Raudaschl
			(3) Italy	Fabio Albarelli
	1972	35	(1) France	Serge Maury
			(2) Greece	Ilias Hatzipavlis
			(3) USSR	Victor Potapov

THE RULES AND AUTHORITIES OF RACING

International organizations

The three international bodies which look after the sport of racing various types of boat are the International Yacht Racing Union (IYRU), the Offshore Rating Council (ORC) and the Union Internationale Motonautique (UIM). The last is the authority for power-boat racing; the others deal with sailing yachts.

The International Yacht Racing Union is the world authority for racing sailing boats. It does not deal with power boats or cruising and other aspects, although in recent years it has encroached on a few general international yachting activities such as international collision regulations and environmental problems concerned with boats on an international basis. Twelve regions of the world are represented on the 'Permanent Committee', which is the central voting body of the IYRU. These regions are the British Commonwealth (except Canada, Australia and New Zealand), Central Europe, East Europe, South Europe, Iberia, Low Countries, North America, Scandinavia, South America, the USSR, East Asia and South-west Pacific. The IYRU recognizes twenty-five classes which are officially designated as 'International'. These contain 4 one-design keel boats, 2 restricted class keel boats, 14 one-design centreboarders, 2 restricted class centreboarders, 1 multihull one-design, 2 multihull restricted classes.

The number of countries with national yachting authorities affiliated to the IYRU is 70 (in 1974). The headquarters of the IYRU has always been combined with that of the British national authority (Royal Yachting Association) in London. The Secretary had always been in the same until 1974 when the jobs were split. The load of business in the IYRU office is, contrary to what might be thought, very much less than the national authority that supports it. This is because the latter has a wide range of responsibilities while the IYRU is restricted in scope to racing and meets only once a year.

The Union has nine committees in addition to the Permanent Committee the names of which give a good idea of its functions. The committees are:

Keel Boat Technical Committee
Centreboard Boat Technical Committee
Constitution Committee
Class Policy and Organization Committee
Multihull Technical Committee
International Measurement Committee
Internation Regulations Committee
Racing Rules Committee
Youth Sailing Committee

The IYRU was the first manifestation of organized international yachting on a regular basis. It was formed at a conference held in Paris in October 1907. There were two main objects in forming the Union. First to unite the various national rules on the way yacht races were conducted – the Racing Rules; secondly to create an international system for a class of yachts – the International Rating Rule. The inaugural members were Austria-Hungary Denmark, Finland, France, Germany, Great Britain, Holland and Belgium, Italy, Norway, Sweden, Switzerland and Spain. The national authority in each of these 11 countries or groups was either a federation (for example the 'Swedish Association of Yachtsmen') or senior club (for example the 'Royal Italian Yacht Club'). The first Secretary was Brooke Heckstall-Smith and the first members of the Permanent Committee were Alfred Benzon of Denmark, M. Le Bret of France, Professor Brusley of Berlin and R. E. Froude of England. The national authorities in the member countries all adopt the IYRU Racing Rules: in Britain they were approved on 4 December 1907 and have been used as amended (not quite out of recognition) ever since. The conferences on yacht measurement were parallel from January 1906 onwards. The USA did not join the IYRU, in any case it had no national authority until 1925. After NAYRU was formed it gradually came to co-operate with the IYRU, but did not join until 1952. By 1936, the Union had 24 member countries.

The British were, therefore, paramount in the conduct of international yachting at that time. Heckstall-Smith gave the following figures for the number and tonnage of sailing yachts in a number of countries in 1906.

Country	**Number of yachts**	**Tons**
United Kingdom	2959	53025
Colonies	322	4563
Germany and Austria	599	8371
France	363	6300
Norway and Sweden	300	4369

Heckstall-Smith said of these figures, 'In these circumstances there can scarcely be any doubt that it is due to England to take the leading position in international yachting. We are sure that other nations will feel much indebted to her for having done so and for having satisfactorily initiated the International Rules of measurement and sailing.'

The object of the Union today is given as 'the promotion of the sport of amateur yacht racing throughout the world without discrimination on grounds of race, religion or political affiliation'.

The authority recognized by the IYRU for the conduct of ocean-racing class yachts is the international Offshore Rating Council (ORC). Its primary task is the administration of the rating rule for ocean racers – the International Offshore Rule (IOR). It has a Technical Committee to handle this and also committees to look after measurement (in order to physically measure yachts to the IOR), safety and special regulations, level rating classes (Ton Cup boats, see p. 43), time allowances and class organization and recommendation. The headquarters of the ORC are in London, but separate from the IYRU.

The ORC was formed as a result of ocean racing yachtsmen asking the organizations which had the most important rating rules to try and unite them. These were the rules of the Cruising Club of America and the Royal Ocean Racing Club (England). A small committee headed by Olin J. Stephens Jr worked on this for two years, starting in 1967. In 1970, the respective American and British clubs allowed their own rules of rating to lapse and began to use the IOR for their races. National authorities and clubs round the world also adopted the rule, which is now the only first-class rule of measurement and rating used for offshore boats. Eleven groups of countries form the council of the ORC, as it only accepts representatives from areas which have ocean-racing yachts. These areas are: Australasia, Benelux, Brazil, West Germany, France/Spain, Italy/Yugoslavia, Greece/Turkey, Japan, Scandinavia, South America (except Brazil), United Kingdom (with Ireland), United States (with Canada/Mexico). The

Tornado the fastest Olympic class

Tornado the fastest Olympic class

The word 'yacht' was first applied to such a craft as this in the 17th century. Elegance has been the characteristic of sailing-yachts ever since

IYRU meets once a year (November) in London (though its committees meet elsewhere at other times); the ORC meets twice a year, in London in November or in one of the member countries. The most important results of its meetings are invariably amendments to the IOR, necessary because of developments in the design and construction of yachts and racing techniques. Other results might be additional safety and related equipment and constructional rules for security when racing and the results of work on the time allowances that are applied to the rating of yachts of various sizes. The first Chairman of the ORC was David Edwards, formerly Commodore of the Royal Ocean Racing Club.

Since its inception, the numbers of yachts rated under the IOR have risen each year. A count is taken each September and in 1974 the figures were:

NAYRU (USA, Canada, Mexico)	4362
UK and Ireland	1834
Italy	415
France	664
Germany	722
Netherlands	524
Belgium	85
Greece	43
Spain	35
Yugoslavia	23
Austria	15
Poland	18
Total Europe (ex-Scandinavia)	4378
Sweden	570
Denmark	161
Finland	107
Norway	69
Total Scandinavia	907
Australia	354
Japan	240
New Zealand	115
Hong Kong	60
Total Far East	769
Argentina	99
Brazil	65
South Africa	47
NAYRU and UK total	6196
Rest of world total	4431
Total rated yachts	10 627

British national authorities

In the United Kingdom the national authority for sailing and power-boating is the Royal Yachting Association. It was formerly the Yacht Racing Association, which was only concerned with the racing of sailing yachts. In the nineteenth century, clubs in Britain each had their own rules of rating, time allowance and racing (collision) rules. Often these were in a very simple form. For instance, the Royal Southern Yacht Club had 12 sailing regulations and a measurement rule of a few lines. 'Owners of yachts which are entered to sail shall draw lots for stations' (it was a standing start). 'Vessels on the larboard tack to give way to those crossing on the starboard tack; and when two vessels by the wind are approaching the shore or mud together, and so near to each other that the leeward one cannot tack clear of the weather-most, and must run on shore by standing further on, such weather-most vessel on being requested, is immediately put about. A vessel by the wind is not to give way to a vessel sailing large.' Of course, several of the clubs had similar rules, following general usages of the sea, such as port tack giving way to starboard, as above and as everywhere today, but differences there were to cause confusion. In order both to unify the racing rules and find common rating rules, in November 1875 the Yacht Racing Association was formed. The leading persons were Count Edmund Batthyany, Captain Hughes and Dixon Kemp. The last, a famous yacht-designer, author and journalist, became the first Secretary. The Marquess of Exeter was elected first President. In 100 years there have only been four secretaries: Dixon Kemp until 1898, Brooke Heckstall-Smith until 1944, Francis Usborne until 1967 and then Nigel Hacking until 1975. At first the Royal Yacht Squadron would not acknowledge the

Dixon Kemp

right of the YRA to legislate for racing rules, but the Prince of Wales (later King Edward VII) became President of the YRA in 1881 and as he was already Commodore of the Royal Yacht Squadron, his diplomacy brought the bodies together.

The steps towards becoming an authority for all aspects of boating took place after 1945. In 1947 because of difficulties such as fuel rationing for boats, timber allocations for building and threats of tax on the purchase of small craft, yachtsmen found that a representative body to talk to Government departments and other official bodies was a necessity. A 'General Purposes Committee' was, therefore, formed to handle matters outside yacht racing. The work in legal, negotiating and administrative business for yachting in general grew so much that in 1952, the YRA became the Yachting Association (YA). In 1953 the royal prefix was added and the Association became the RYA. It has corporate (club and associations, sail and power) members and individual members in various categories.

The biggest membership of the RYA is invariably the latest figures at any time. Progress of membership at certain stages was as follows:

Year (31 December)	Club membership	Personal membership
1935	264	231
1946	321	273
1954	714	2380
1964	1271	19 178
1974	1600	41 000

The work of the RYA is very large indeed and can best be summarized by listing its organization and purpose of various committees. There is an elected council with a President Chairman and Deputy Chairman. The permanent paid staff include as well as the Secretary, Training Manager, Cruising and General Services Manager, Development Manager, Racing Manager, Cruising Secretary, Assistant Racing Manager, Chief Measurer, Olympic Appeal Manager all with numerous supporting staff.

There are 'divisional committees for the various aspects of boating with sub-committees'.

Divisional	Sub-committees
Central Management	
Central Finance	
Development	Membership
Training	
General Purposes	Cruising
Yacht Racing	Centreboard Cruiser Racing Advisory International Racing Keel Boat Ladies' Championship National Measurement Racing Rules Team Championship World Sailing Speed Record Olympic British Olympic Yachting Appeal
Power-boat	Offshore Racing Sportsboat Racing Hydroplane Racing Competition Cruising Inflatable Boat Racing Technical

The most expensive survey ever undertaken by a non-commercial body in the interests of yachtsmen was commissioned by the RYA in July 1972. This was a study of its aims, organization and development and was carried out by management consultants, Spencer Stuart Corplan of London at a price of £15 000. The resulting report (published in March 1973) has given the most significant estimate of what constitutes the boating public. Because it is leisure orientated, because there are no comprehensive boat registrations, and because of its wide diversity, no other systematic impartial surveys have been available: it was known only that there had been very great growth in boating in Britain since 1945.

In 1972, according to the report, well over 2 000 000 people in the United Kingdom have some interest in boating. However the majority of these (1 250 000) have no significant commitment to boating as a hobby. They might read with interest a newspaper item about, say, the Admiral's Cup, but would not take a yachting magazine: they might go afloat with friends, if asked, or hire a boat for a holiday.

The remaining 750 000 were said to be divided as follows:

Power-boat racing	2000
Offshore and similar sail-boat racing	15 000
Offshore and coastal cruising	20 000
Racing restricted and one-design classes, mainly dinghies	200 000

The other 500 000 are interested in local and estuarial sea cruising, local day and week-end sailing, day cruising in trailed or locally parked run-abouts, or are potterers and novices. Among the large number of dinghy racers in the fourth category are some who race regularly but others who are only active a few days each year.

The objectives for the 1970s for the RYA have been laid down with particular clarity. In general they are to serve the needs of all who go afloat for recreation or sport under sail or power (this excludes rowers and fishermen for instance). Its specific tasks are *representation* with all sorts of official bodies and other organizations, local, national, international; *racing* affairs, in which it acts as governing body in Britain for both sail and power; *training*, in which it encourages certain standards and a scheme for attaining them. Less well-defined objectives are *facilities* where these are appropriate and the circulation of *information* to and among yachtsmen.

The direct appeal for substantial funds by the RYA comes from its British Olympic Yachting Appeal which culminates just before each Olympic Games. The purpose is to raise money to supplement Government grants in order to select, train and support the Olympic yachting team. The maximum sum raised was £80 000 in the four years before the last Olympic Games in 1972.

The largest part of yachting in which the RYA does not take a direct part is in offshore racing. It generally leaves guidance of this to the Royal Ocean Racing Club, although remaining the ultimate national authority in Great Britain for the sport. The reason for this is the unique position in the world of the RORC as a club organizing ocean racing. The RORC gives more races of 200 miles and over every season than any other club in the world. Thus it has built up an exceptional expertise on national and international level. The number of races in its 50th anniversary year, 1975, was 23 and their total mileage was 5349. No other club in the world organizes such a string of events. (RORC assistance with the Round the World race is not included.)

The North American Yacht Racing Union was the national authority for the racing of sailing yachts in the United States, Canada and Mexico. Canada and Mexico have their own national authorities as well, but NAYRU was established earlier and is responsible over all for many aspects outside the strictly national ones such as grants to their own sailors by the Government of Canada. However in 1975, after 50 years, each country reverted to its own national authority and the United States Yacht Racing Union was formed. The Canadian Yachting Association became autonomous.

American national authorities

The first attempt to form a North American Yacht Racing Union, for the customary reasons to unify racing and measurement rules, was on 30 October 1897, when a meeting was held at the Fifth Avenue Hotel, New York. A hundred and eight clubs elected to form the Union including the Lake Yachting Association, the Yacht Racing Association of Long Island Sound, Pacific Inter-club Yacht Racing Association and Royal St Lawrence Yacht Club. Oliver Cromwell was elected Chairman and W. P. Stephens, Secretary. However the Union did not get any further because of the opposition of the New York Yacht Club which controlled all large yacht racing and did not wish to relinquish its authority. In this it was supported by some of the leading New England clubs – an exact repetition of the early situation between the RYS and YRA, except in the American case, the national body collapsed.

A second and successful attempt was made in 1925 and NAYRU has been the acknowledged authority for yacht racing since then; except by the New York Yacht Club until 1942 and the important Eastern Yacht Club (at Marblehead, Mass.) until 1946. Even today the NYYC holds the America's Cup under the 'rules of the NYYC', which are in practice the NAYRU Racing Rules. The leading light in the formation of NAYRU was Clifford D. Mallory of the Indian Harbor Yacht Club, who became its first President. After its formation NAYRU immediately started to put out international feelers, indeed this was one of the reasons for unifying American yachtsmen in this way. This was the time when the British-American Cup races (see p. 69) were gaining support – an incentive to transatlantic links.

In 1926 Clifford Mallory and Clinton Crane went to London as observers of the IYRU meetings: as a result, NAYRU in 1929 adopted the Racing Rules of the IYRU, which by then had voluntarily incorporated some NAYRU suggestions. From then one set of Racing (right of way) Rules was used all over the world, though coupled with local prescriptions – as they are today. But in 1948, NAYRU adopted its own set of racing rules, known as 'Vanderbilt Rules' after their originator. These were quite different in respect of right of way, as well as other details. In 1952, NAYRU at last joined the IYRU. The IYRU then made trials of the Vanderbilt Rules, which were adopted in principle. By 1960 there were then one world set of racing rules – the present IYRU Rules (as amended since) – based on the NAYRU system. One more effect of NAYRU and IYRU joining together was the subsequent recognition of effective International class organizations.

In 1970 when the Cruising Club of America and the Royal Ocean Racing Club had amalgamated their rating rules into the IOR, NAYRU took over from the CCA as the administering authority for the IOR in North America and became responsible for sending delegates to the ORC as well as the IYRU. Meetings of the ORC are held in the USA from time to time. In 1971 there was a meeting in San Francisco; in 1974 there was one in Marblehead.

In 1974 the USYRU had the greatest number of individual members in its history, 9134 (to March 1974), while the yacht club and class association membership was 658. In addition whole regions are affiliated and they have their own club membership. These regions such as the Florida Sailing Association, and the Dixie Inland, Great South Bay, Maine and Nova Scotia Yacht Racing Associations numbered 36.

The work of USYRU as shown by its committees is divided into responsibilities by these groups of committees:

1. *Officers and Executive Committee*
2. *Special committees*
   a. Racing Rules
   b. Life-jackets
   c. Portsmouth Numbers
   d. Amateur Eligibility
   e. Race Committee Seminar

3. *Class racing committees*
   a. Class Racing
   b. US Olympic Yachting
   c. Class Associations
   d. Junior Activities
   e. Junior North American Championships
   f. Women's Sailing
   g. North American Championships
   h. Single-handed Championship
   i. Inter-club Race Championship

4. *Offshore racing committees*
   a. Offshore Racing
   b. Administrative
   c. Technical
   d. Rating Rule
   e. Level Rating Classes
   f. Safety at Sea

There are two other groups of people: Firstly, those who serve on the various committees of the IYRU, 21 of them (every IYRU committee has at least one American on it). Secondly, those who are nominated for the ORC and its committees, 13 of them. Henry S. Morgan is a Councillor of Honour on the IYRU and Olin J. Stephens has this distinction in the ORC. In both cases this is in recognition of their great services in bringing the American and European interests together in these international bodies.

Both the RYA and USYRU give awards to individuals for services to their countries in the cause of the sport. The USYRU Nathanael G. Herreshoff Trophy was donated by the National Association of Engine and Boat Manufacturers and is given annually to the individual who has contributed most to the sport of sailing in North America. The award has been made since 1957.

1957	Henry Sears
1958	Henry S. Morgan
1960	George D. O'Day
1961	James M. Trenary
1962	Julian K. Roosevelt
1963	R. S. Stevenson, OBE
1964	Olin J. Stephens II
1965	Leonard Munn Fowle
1966	Allegra Kanpp Mertz
1967	Everett B. Morris
1968	Harold Stirling Vanderbilt
1969	Paul H. Smart
1970	F. Gregg Bemis
1971	J. Amory Jeffries
1972	Harry C. Melges Jr
1973	George R. Hinman
1974	William S. Cox

The RYA Award scheme was introduced in 1971 for those who have made an outstanding contribution in Great Britain to 'popular pleasure boating'. There is no regular period for presentation, the Award consists of a scroll and lapel badge and the RYA Council decides on suitable persons from time to time. Up to 1975, the RYA Award-winners had been:

Captain John Illingworth, RN (ret.)
Sinbad Zillwood Milledge
Hon. Ewen E. S. Montagu, CBE, QC, DL
Major E. G. M. Pearce, RE
Gerald Sambrooke Sturgess
Captain Terence W. B. Shaw, DSC, RN (ret.)
Miss Alex Cowie
J. D. Slater
Major-General Ralph H. Farrant, CB
Group Captain E. F. Haylock, RAF (ret.)
Vernon Stratton

A note about the basic right of way rules
Between 1875 and 1960 the IYRU rules followed the basic laws of the sea in respect of sailing ships meeting. This was that a vessel running free gave way to one which was close-hauled. It had been customary for centuries, since sailing close-hauled was very difficult for square-riggers and all sailing craft. If both vessels were close-hauled or both were running free then the yacht on starboard tack had right of way. The Vanderbilt Rules changed this. Instead, starboard tack had right of way in all circumstances. It changed the old practice because now a port-tack boat running free had to give way to a starboard-tack one close-hauled. This was one of the basic changes incorporated when the NAYRU and IYRU Rules combined. Of significance was the result that when the international regulations for prevention of collision at sea were next revised by all the Governments concerned with all ships on the high seas, the rules for sailing vessels were brought into line with this new basic yacht-racing right of way system.

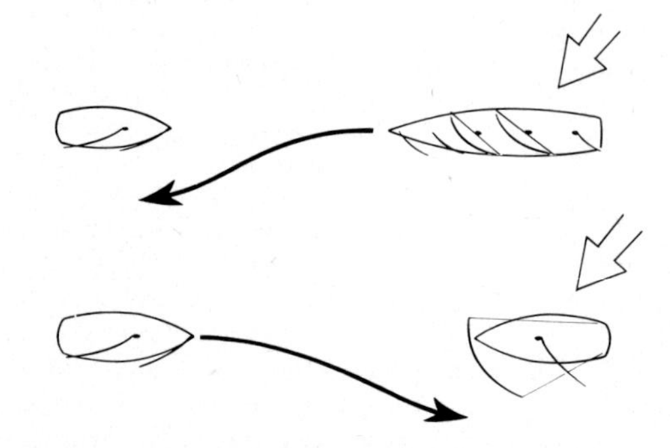

Till 1960, a vessel close-hauled on the wind always had right of way over one reaching or running.

Since then a sailing boat with the wind on the starboard side always has right of way over one with the wind on the port side, regardless of how the sails are sheeted in.

TIME ALLOWANCES AND HANDICAPPING

The unsolved problem in yacht racing is to find a satisfactory result, when yachts of different sizes and types race together. Where size is similar, but shape is different, yet they are designed to the same rating, the first boat home is the winner. And in one-design racing the results are the same as the order in crossing the line: but where the ratings are different, or the boats are not rated but are diverse in ability, the customary system is to correct the finishing order by application of a *time allowance*. This is widely known, but less accurately, as 'handicapping'.

No other sport gives a time allowance in this way, after the race is complete, but other methods were tried many years ago and discarded. For instance in horse racing jockeys have weights added as handicap. In yachting, this can happen in one-design dinghies to bring the boat into the correct weight under its particular rule, but not in 'handicapping'! Giving a 'start', by letting one yacht cross the line earlier than another is used exceptionally in 'pursuit' races. Usually these are not of a serious nature. However in the 1973–4 Whitbread Round the World race, the yachts were started at intervals of hours or days, on the last leg from Rio to Portsmouth. This was based on the performance in the previous legs of the course and was intended to bring them back to England at approximately the same time. The course from Rio to Portsmouth is, therefore, the longest pursuit race ever held. (For this time allowances were also applied to the elapsed time of each yacht. A true pursuit race, if defined, would apply the whole time allowance at the start-line, so that the first yacht to finish was then the winner.)

Early time allowances

The first attempt to 'allow time' to yachts in a race because of differing sizes was the Royal Yacht Squadron regatta in 1829. As we have recounted already, yacht racing was originally match racing. So one owner challenged that his yacht was faster than another. Probably a challenge did not materialize unless the yachts were thought to be of about equal speed, indeed this was what was in dispute. But once this extended to several yachts racing for a prize, the largest yacht invariably won. To win a race it was first necessary to build a yacht larger than those of the other owners. There were cases of owners lengthening their yachts during the winter. In 1827 Lord Belfast inserted a new midship portion into his yacht *Louisa* in this way.

In 1829, the Squadron yachts ranging from 40 to 140 tons by the Government tonnage measurement of the day, were divided into 6 classes. Each group was started after the smaller one had covered a certain distance, in proportion to the distance of the course. The 1st class gave ½ mile to the 2nd, 1¼ miles to the 3rd and 2¼ miles to the 4th. For the first time in history class prizes were allotted. In 1831, King William IV, asked that 'His Majesty's Cup' for the annual regatta, be given on the basis of this time allowance. This royal approval of time allowances marked the end of the complete big boat dominance of the 1820s.

The first actual application of time allowance was in 1838, when the yachts in the RYS regatta were given 3 minutes for every 10 tons at the start. This was run as a pursuit race. The system did not last long, partly because the racing was from a 'standing start' at anchor.

The first application of time allowance in the modern manner was by the RYS in 1841. The yachts started together, but the elapsed time of each yacht was corrected by 1 second per mile per ton. (So a yacht of 50 tons racing over a course of 20 miles against a yacht of 75 tons would have 500 seconds – 8 minutes 20 seconds substracted from her time! After this the vessel with the shortest time would be the winner.)

The 217 ton schooner *Dolphin*, a Royal Yacht Squadron of 1837. She was owned by G. H. Ackers who introduced systematic time allowances into the regatta at Cowes.

The first established time allowance system intended to be used for all yacht races was invented in 1843 by Mr George Holland Ackers, a member of the Squadron and owner of several yachts at different times, including the 217 ton schooner *Dolphin*. He called his time-scale 'Acker's Graduated Scale'. It was graduated because the allowance was not a linear value for each successive ton, but reduced in time per mile as the tonnage increased. In other words the allowance between a yacht of 60 and another of 55 tons would be less than the allowance between yachts of 45 and 35 tons. This again was more in line with modern practice where a formula achieves this sort of sliding effect. Acker's scale also made allowances for different rigs, as well as tonnage, credit being given for schooners and square-riggers against cutters.

The first time allowance system in the United States was used by the New York Yacht Club at its very first regatta on 16 July 1845. Tonnage was by US Custom House measurement and vessels were allowed 45 seconds per ton for the course. There was no allowance for rig, although there were sloops and six schooners in this historic race which was "from a stake boat off Robbin's Reef, to a stake boat off Stapleton, Staten Island, thence to and around the south west Spit buoy, returning over the same course'.

The first division by classes and form of graduated scale in the USA was in the NYYC regatta of 1847. There were three divisions: two by size and the third for non-club boats, probably mainly working sailing-boats of fast type. The scale was 35 seconds per ton for the big class, 45 seconds per ton for the smaller class and 40 seconds per ton for the 'outsiders'.

In 1871 the Seawanhaka Corinthian Yacht Club of Oyster Bay on Long Island Sound formulated a time allowance system based on the length. It is undisputed that speed of displacement vessels is proportional to the square root of the effective length (however that may be arrived at) or rating, which is usually rated length (see Section 5). The time allowance per mile in seconds between two boats of Lengths L and l was therefore given as

$$0{\cdot}4\left(\frac{3600}{\sqrt{l}}-\frac{3600}{\sqrt{L}}\right)$$

The 0·4 was put in because the generally light winds of Long Island Sound seldom allowed yachts to attain their full potential and this allowed a proportion of the potential speed. After a few years this was not considered satisfactory because in 1875, the Seawanhaka set up a committee 'to revise the system of time allowance'. This is the first recorded attempt by a club to progressively improve the method of allotting time allowance in its races.

The first attempt by the Yacht Racing Association in Britain to introduce a time allowance based on *rating* was in 1886. The rating was found from the length and sail area rule of the time. The time allowance per mile in seconds between two boats of ratings R and r was $\sqrt[5]{R}-\sqrt[5]{r}$.

The first international time allowance system was introduced by the International Yacht Racing Union in January 1908. The system was elaborate and a compromise to suit the various European countries which had been involved. In Scandinavia there were four time-scales depending on the wind during the race (how this was measured and decided is not clear), but in Britain only the 'medium' scale was ever used. There were different factors for different sizes and rigs. The allowances were by time-on-distance – so many seconds per mile. In the medium scale for cutters, examples of the allowances were that a 5-metre (rating) was given 50 seconds per mile by a 6-metre, 122 seconds per mile by an 8-metre, 205 seconds per mile by a 12-metre and 307 seconds per mile by a 23-metre.

On the light wind scale, by comparison, the 5-metre was given only 84 seconds by an 8-metre.

Later systems

The oldest time allowance system still in widespread use is the time-on-distance tables of the New York Yacht Club which originated in 1908 and are still used as 'The NAYRU time allowance tables' today. As now laid out, they show how many seconds per mile a boat of a certain rating gives to a boat of another rating. They were based on the weather experienced in the north-east United States in the sailing season, where light winds predominate. Based on the established square root of the rating, the system assumes that in these winds and round a triangular course of close-hauled, running and reaching, the boats of different rating draw away from one another at a proportion of 0·6 of the full potential. This gives the time allowance per mile in seconds as:

$$\frac{2160}{\sqrt{r}} - \frac{2160}{\sqrt{R}}$$

2160 is 0·6 of the number of seconds in an hour: the form is, therefore, the same as the Seawanhaka time allowances of 1871. The NAYRU time allowance system has been used with ratings since 1908, though these actual ratings have been arrived at under quite different rating rules. These rules (Section 5) have included the Universal Rule, the Cruising Club of America Rule and the present IOR. Provided there is a triangular course, or equivalent due to changes of wind direction, and a wind between 8 and 25 knots the NAYRU time allowance tables work well. In extreme conditions, such as the wind in one direction relative to the course throughout a race, or very light or very strong winds then results are weighted in favour of particular sizes of yacht. This is a defect in all time allowance systems.

The last time-scale to be recommended officially for international use was the IYRU time allowance scale of 1935 for use with the International (metre-boat) Rule. The scale was numerically close to the NAYRU one. A speed of so many seconds per mile was given as the theoretical speed of a yacht of given rating. To find the allowance of one yacht against another, the difference of the theoretical speeds were taken. For instance a 6-metre had a 'speed' of 519 seconds per mile; an 8-metre had 452 seconds. The allowance per mile which an 8-metre gave to a 6-metre was, therefore, 67 seconds. No international time-scale now exists.

The most frank admission by an international body responsible for time allowances was by the Offshore Rating Council in April 1973, when its Time Allowance Committee stated that it could not find any time allowance for use with the IOR which it could recommend internationally. It asked national authorities to continue experimenting with different time-scales and meanwhile it would try and build up a data library of race results as a basis for attempting to find a system which could eventually be recommended. This remains the current situation.

RORC systems

The most flexible organization when deciding time allowances has been the Royal Ocean Racing Club. In its early races from 1925 to 1935, a time-on-distance method similar to the Americans was used. However the full potential of the rating was used and the time allowance per mile in seconds between two boats of rating R and r was:

$$\frac{3600}{\sqrt{r}} - \frac{3600}{\sqrt{R}}$$

Because it was seen that the boats seldom sailed at full potential the allowance was changed in 1934 to:

$$0{\cdot}8\left(\frac{3600}{\surd r}-\frac{3600}{\surd R}\right)$$

However in the 1935 race from Burnham-on-Crouch to Heligoland, in a very strong south-west wind the race was won by the smallest yacht a double-ended ketch, *Hajo*, despite the fact that the winning yacht the German *Asta* finished first at an average speed over the 910 miles of 10·8 knots. On the time-scale it worked out that *Asta* would have had to sail 13 knots to beat *Hajo*, an impossibility. The anomaly was due to the time-scale.

The RORC, therefore, introduced a new time allowance system, the time-on-time system. Instead of a yacht being allowed so many seconds per mile against an opponent, a smaller yacht was allowed so many seconds per hour (or fractions of an hour). Therefore if the race was a fast one, the smaller yacht would not get so much allowance as in a slow one, although the distance was the same. The big disadvantage in this method is that in flat calms or near calms when the yachts of all sizes are hardly moving in relation to one another, the smaller yacht is amassing time against the larger without effort. As this situation is less frequent around the British Isles than on the eastern seaboard of the USA, it has been an acceptable deficiency in time-on-time systems. Time-on-time is scarcely known in North America.

The first time-on-time method, started by the RORC in 1936, took this form. Each yacht was allotted a TCF derived from its rating:

$$\text{TCF} = \frac{\surd(\text{Rating}) + 2}{10}$$

After the completion of the race, the yachts elapsed time is multiplied by the TCF to give a corrected time. The yacht with the best (shortest) corrected time is the winner. For instance if the yacht's time over the course was 8 hours 30 minutes and her TCF was 0·8525, then her corrected time would be 7 hours 14 minutes 46 seconds.

The first alteration to the time-on-time formula by the RORC was in the early 1950s when it was found that the smallest yachts in the fleet predominated in most races. It was changed to

$$\text{TCF} = \frac{\surd(\text{Rating}) + 3}{10}$$

When the IOR was adopted by the RORC in 1970, the formula was again changed to bring the corrected time figures approximately in line with the previous scale numerically: there was no change of principle or intended allowance:

$$\text{TCF} = \frac{\surd(\text{Rating}) + 2{\cdot}6}{10}$$

The first attempt to combine time-on-time and time-on-distance was also made by the RORC for use in its races starting in 1972.

No other clubs adopted this and because of its lack of simplicity and the fact that it still did not solve the problem of extreme conditions, it was phased out after the 1974 season. It took this form: of a 'yacht rating factor' deduced from the rating of the yacht:

$$\text{YRF} = \frac{\dfrac{\sqrt{R}}{\sqrt{29}} - 1{\cdot}0}{\dfrac{1{\cdot}5}{2{\cdot}75}\sqrt{R} - 1{\cdot}0}$$

Each race had a nominated course distance, from which was obtained a distance factor

$$\text{DF} = \frac{\text{course distance}}{2{\cdot}75}$$

Then if ET is the elapsed time, then the corrected time of each yacht was found after she had finished by:

$$\text{CT} = \text{ET} + \text{YRF}(\text{DF} - \text{ET})$$

For the 1975 season, the RORC abandoned this system of combined time and distance (for some unknown reason called the 'performance factor') and reverted to a time-on-time method – the fifth change in its history. The 1975 formula was:

$$\text{Time multiplying factor (TMF)} = \frac{0{\cdot}2424\sqrt{R}}{1 + 0{\cdot}567\sqrt{R}}$$

TMF is multiplied by elapsed time to give corrected time. This meant that a yacht of 29 ft rating was scratch boat and its corrected time always equalled its elapsed time. Larger boats would have a corrected time bigger than their elapsed time and boats smaller than 29 ft rating always in a race had a corrected time numerically smaller than the elapsed time. The only object of this was to show the corrected times numerically close to the elapsed times, which yachtsmen were thought to prefer. The yacht with the shortest corrected time is still the winner.

The most misunderstood part of the results of any handicap race (or race on time allowance) is the corrected time. The corrected time of each yacht is only there to deduce the finishing order. The actual figure entirely depends on the formula or time allowance system used. Therefore there is no such thing as record corrected times for a course. If yacht A beats yacht B by, say, 15 minutes on *corrected time*, the actual number of minutes by which B should have improved her performance will depend on the system. If B is bigger than A she may well have finished ahead of her (and so will in this case need to finish even farther ahead to beat or equal A's corrected time).

The last occasion that time allowances were used in an America's Cup series was in July 1920 in the *Shamrock IV* versus *Resolute* match. The defender *Resolute* was allowed between 6 minutes 40 seconds and 7 minutes 1 second deducted from her time (depending on the course laid). After 1920, the yachts have always raced level with the same rating.

The most inexpensive, but also arbitrary form of handicapping, is by application of a figure for time allowance by a club handicapper. This procedure, which in effect penalizes yachts which perform well, is closer to the correct use of the term 'handicapping'. Handicapping implies, as in horse racing, the function of a handicapper, who endeavours to equalize the chances. Generally when a yacht is new or arrives for a race, the handicapper allots a figure, either a TCF or so many seconds per mile allowance against the scratch yacht. After one or more races, this can be adjusted on the form the yacht shows. Different clubs all over the world have their own ways of going about this. Such procedures work satisfactorily at club level and where the yacht-owners are of comparable ability and away from the intensity of national or international competition.

The main problem lies in the judgment of the handicapper, but the satisfaction with results depends on the attitude among the competitors. They may be pleased if the corrected times of most of the yachts, especially the leading ones, are close; however the closeness of results after correction, contrary to widespread belief, does not mean that the method used is a good one. For the 'true' result, as in a one-design class, may involve the boats being some minutes apart. A subjective judgment of the results is, therefore, necessary for club handicapping, and the handicapper will be influenced by the type of result the club members and competitors assume is fair.

Portsmouth Yardstick

The only widely accepted system of handicapping on performance (as opposed to obtaining a rating by measurement and then working a time allowance) is the Portsmouth Yardstick scheme. This was devised by Sinbad Zillwood Milledge of Chichester, England and was first used by the clubs of Portsmouth Harbour for running races for centreboarders, where they were not of the same class. The scheme has now spread all over the world including the United States. It has also spread from dinghies to its use by small cruisers.

The Portsmouth Yardstick has developed in effectiveness with use over the years and there is nothing similar to it. The basis of the scheme is essentially one of *class*, so that a dinghy or any controlled class has a Portsmouth number. Any boat of that class turning up for a race at a strange club can be given that number as a rating of its performance. A problem arises with cruisers which are of a 'commercial class', but are in fact far from standard. The RYA has, however, defined what is standard for a sailing-cruiser of such a class and given guidance to enable handicappers to vary the number depending on whether the yacht has, for instance, a single keel rather than twin, a gaff-sail rather than a Bermuda, or no engine fitted when with other boats of the class it is standard.

The Portsmouth number itself is a time over a common but unspecified distance. It supposes that a Fireball with a number of 85 covers a certain distance in 85 minutes, while a Mirror dinghy with a number of 122 will, in the same race, take 122 minutes over the same distance. The numbers run from 60 upwards – the lower the number the faster the boat. The fastest class with a primary yardstick is the International Tornado catamaran. The slowest boat with a secondary yardstick is the International Optimist.

The definitions used in the scheme are considered particularly important because there are different grades of number. Yardstick numbers are agreed by the national authority, or several clubs, for a class boat (one-design or commercial class). The number represents the average performance of the class and not the fastest boat in it. *Primary yardsticks* are well attested and for all practical purposes do not change, but change is possible if new information comes to light and is passed on to the national authority running the scheme. *Secondary yardsticks* are still open to correction and clubs sometimes vary them, but they are fairly close to the right number, and are used if other data is not available. Provisional *Portsmouth numbers* are allotted to boats where there is no previous information on them. An important point about the Portsmouth scheme is that any boat (even if not in a class) can sail in the race when she may never have had a number. The handicapper must allot her one which is subsequently adjusted on performance. This is quite a different situation to the use of measured ratings. It is only by racing, and by club handicappers sending in data to the national authority, that primary and secondary yardsticks can be built up. These can then be published and used by all clubs as boats, which have hitherto not raced with them, arrive to sail in handicap events.

A number of ready reckoner type tables are available to help clubs work out corrected times direct from Portsmouth numbers of the competitors and their elapsed times. They also enable handicappers to allot new numbers from observed performance. They are the Langstone Tables compiled by Milledge, Peggs TCF Tables, Bowen Tables, Factorgesima Time Multiplication Tables.

The maximum number of classes of boat with primary and secondary yardsticks accumulated by the RYA up to early 1974 was:

Type of boat	Primary yardstick	Secondary yardstick
Centreboarders	40	58
Keel boats	2	10
Sailing cruisers	17	89
Multihulls	3	7

However sailing-cruisers, because they have not been using the Portsmouth scheme for so many years as one-designs, had 100 classes of boat recorded with provisional Portsmouth numbers.

CHRONOLOGY OF SOME MILESTONES IN YACHTING

Early seventeenth century. Yachts built in the Netherlands: sailed at festivals and by wealthy citizens.

1660 Charles II receives gift of yacht *Mary* from the Netherlands.

1661 First yacht race in England from Greenwich to Gravesend, between Charles II and Duke of York.

1720 The Water Club of the Harbour of Cork in existence.

1772 Starcross Club regatta (England).

1775 Cumberland Fleet race on the Thames.

1815 'The Yacht Club' formed to race at Cowes.

1832 Royal Swedish Yacht Club formed.

1844 New York Yacht Club formed.

1850 Macmullen cruises in *Leo*, 3 tons.

1851 *America* wins the Hundred Guinea Cup at Cowes.

1864 First motor-boat with internal-combustion engine.

1866 (December) First transatlantic race.

1875 Yacht Racing Association formed.

1876 *Sunbeam*'s cruise round the world.

1876 First single-handed crossing of Atlantic.

1886 Tonnage rating abolished in Britain.

1887 First one-design class: Dublin Bay *Water Wags*.

1893 First British one-design class starts.

1893 More big racing cutters launched in Britain and the USA in a single season than ever before or since (British) *Britannia*, *Satanita*, *Valkyrie*, *Calluna*, (American) *Vigilant*, *Navahoe*, *Colonia*, *Jubilee* and *Pilgrim*.

1898 First single-handed voyage round the world completed by Joshua Slocum.

1900 First yachting events in Olympic Games.

1900 First US one-design class starts.

1903 Biggest ever cutter built (*Reliance*).

1903 First international speed-boat race.

1904 First offshore race for amateur manned small yachts: Brooklyn to Marblehead.

1905 Motor boat race held at Monaco.

1906 First transpacific race.

1907 International Yacht Racing Union is formed and international (European) rating rule.

1912 First power boat goes over 50 mph.

1920 First British-America Cup in 6-metres.

1923 First New York–Bermuda race under auspices of Cruising Club of America.

1923 First British national dinghy class (14-footers) formed.

1925 North American Yacht Racing Union formed.

1925 First Fastnet race, seven yachts started.

1931 First power boat goes over 100 mph.

1932 First one-design keel boat (Star) in Olympic Games.

1937 Last year of the big cutters in America's Cup.

1941 First Southern Ocean Racing Conference (Florida) series.

1946 First British national dinghy one-design class (Firefly).

1947 Royal Ocean Racing Club admits boats as small as LWL 24 ft to its races.

1949 North American Yacht Racing Union joins International Yacht Racing Union and one code of racing rules becomes used in the world.

1952 First woman single-hander crosses an ocean.

1953 Smallest ever boat wins Fastnet race (LWL 24 ft).

1955 Water speed of over 200 mph achieved.

1956 First Olympic Games with a two-man dinghy class.

1956 First America's Cup since 1937 (in 12-metre yachts).

1957 First Admiral's Cup race starts new inshore/offshore series concept.

1959 Fastnet race entries exceed 50 boats.

1960 First single-handed transatlantic race.

1962 Australia's first challenge for America's Cup.

1962 First 'Little America's Cup' for catamarans.

1963 Fastnet race entries exceed 100 boats.

1968 Smallest ever sailing yacht (5 ft 11½ in) crosses an ocean (Atlantic).

1969 First single-handed voyage round world without any port of call by Robin Knox-Johnston.

1970 International Offshore Rule is introduced throughout the world, instead of separate British and American rules.

1971 International Star class one-design is 60 years old.

1971 Fastnet race entries exceed 200.

1972 First Olympic Games without any 'formula class'.

1973 Number competing in a single dinghy class championship reaches 230.

1974 First race round the world for fully crewed ocean racing yachts is completed.

1974 International offshore multihull rule is introduced.

1975 First race across the Atlantic for fully crewed multihulls begins.

All these milestones are narrated in more detail in other pages.

3 The America's Cup

The America's Cup story is an historical exercise in exaggeration. Every important aspect of it is more expensive yet more rewarding, more acrimonious yet more demanding of sportsmanship, than any other aspect of yachting. More technological effort is injected into preparations of the contenders than in any other sailing boats, but as a spectacle of sport it is one of the most tedious. It remains the most important of all yacht racing events. It is the oldest international trophy in any sport: the first contest was in 1851. It has the longest unbroken sequence of wins by a single nation or club in any sport.

The America's Cup, symbol of yacht-racing supremacy.

The cup itself is of silver, weighs 8 lb 6 oz and is 2 ft 3 in high. It stands in a case made of glass, but in the middle of the trophy room of the New York Yacht Club, a building of Victorian design at 37 West 44th Street. The cup is fastened securely through to the floor and can be seen on all sides through the case. No one even pretends it is beautiful. After one and a quarter centuries, the members of the NYYC are bound to ensure that it never moves from this (or a comparable) position. The cup was made by R. & G. Garrard of London. The name 'Auld Mug' was coined by Sir Thomas Lipton, five times challenger for the cup. The name comes from the first yacht to win it, *America*, but conveniently symbolizes the supremacy of America in speed under sail and yacht racing. In origin the cup was one of a number given every year by the Royal Yacht Squadron: its original and only name was 'The Hundred Guinea Cup'. This simply described its value as an outright prize, which at that time it was the custom to give. It was, therefore, a valuable piece of silver even in 1851. It was a common fallacy in the nineteenth century that it had been presented by Queen Victoria and was often quite incorrectly referred to as 'The Queen's Cup'.

That the America's Cup is the most one-sided contest in sport is shown by the fact that between 1851 and 1974, the NYYC defender has won on all 22 occasions. Even in the individual races which comprise a cup series, a challenging yacht has won only 7 times out of 77 races sailed. In 2 more races the challenger appeared to have won but was subsequently disqualified for rule infringement.

The primary reason for the invariable defeat of the challenger is that the appallingly expensive operation of mounting a challenge is often taken on by a man with limited experience of yacht racing, who has been successful in business and who is prepared to expend vast sums of money. Often only one yacht has been built for the challenge, so that the challenger is bought rather than selected. Victory, however, has to be vied for and cannot be bought. The motives of a number of the challengers have been mixed. Sometimes they have wished to promote their business through the publicity of the America's Cup races.

The success of the defence lies basically in the attitude of the New York Yacht Club. The Cup is the symbol of American prowess under sail, especially with regard to speed – a tradition going back to the days of America's fast clipper ships and eighteenth-century fast coastal craft.

This attitude is epitomized in the model room of the NYYC – the world's most impressive shrine to yacht racing. In this hall with stained-glass windows are large models of every America's Cup defender and challenger. The walls are covered with hundreds of other half models of yachts which have won races organized by the club. There is no comparison to this room anywhere else in the world.

The NYYC has almost always run trials for several contenders for the honour of defending the Cup and this brings the defender to a higher state of training and tune than the challenger. The defence contenders have, with a few exceptions, been owned by syndicates of men in American business (who are already members of the NYYC), so they have also been able to overwhelm the challenger financially. In addition to all this the United States has a greater number of skilled yachtsmen to draw on than any other country and many who are used to racing in large yachts.

The world's first international exhibition – the Great Exhibition of 1851 at the Crystal Palace – was the occasion for asking the United States to send one of its reputedly fast schooners to England. The promoters of the exhibition thought that an American yacht might compete in various races in the Solent that were being held as 'an auxiliary feature' of the exhibition. The invitation to *America* to compete in Royal Yacht Squadron regattas was contained in a charming letter from the Commodore, the Earl of Wilton, to John C. Stevens, Commodore of the New York Yacht Club. It said:

7 Grosvenor Square,
London.
February 22 1851.

Sir:—Understanding from Sir H. Bulwer that a few of the members of the New York Yacht Club are building a schooner which it is their intention to bring over to England this summer, I have taken the liberty of writing to you in your capacity of Commodore to request you to convey to those members, and any friends that may accompany them on board the yacht, an invitation on the part of myself and the members of the Royal Yacht Squadron, to become visitors of the Club House at Cowes during their stay in England.

For myself I may be permitted to say that I shall have great pleasure in extending to your countrymen any civility that lies in my power, and shall be glad to avail myself of any improvements in shipbuilding that the industry and skill of your nation have enabled you to elaborate.

I remain, Sir, your obedient servant,

Wilton
Commodore of the Royal Yacht Squadron

The schooner *America*, the most remarkable yacht even in her own time, was built in New York by George Steers and William H. Brown, launched on 3 May 1851 and purchased outright by the syndicate headed by John C. Stevens consisting of his brother and four other members of the NYYC. Her dimensions were LOA 101 ft 9 in; LWL 90 ft 3 in; beam 22 ft 6 in; draught 11 ft 6 in; sail area 5263 ft^2. Tonnage according to the measurement of the time was 170. She cost $20 000. After the Cup race of 1851, the yacht was sold to Lord de Blaquiere in England. She cruised and raced occasionally and was sold back to the USA during the Civil War to become a blockade runner. She later became the property of the US Naval Academy and joined in the 1870 America's Cup match. At one time she was refurbished, but by the 1940s was in a poor state of repair. In 1945 the roof of the shed at Annapolis in which *America* was hauled out, collapsed and she was destroyed.

Courageous, winner of every race in the 1974 America's Cup

Southern Cross (yellow) the Australian challenger of 1974 splits tacks with *Courageous*, the American defender

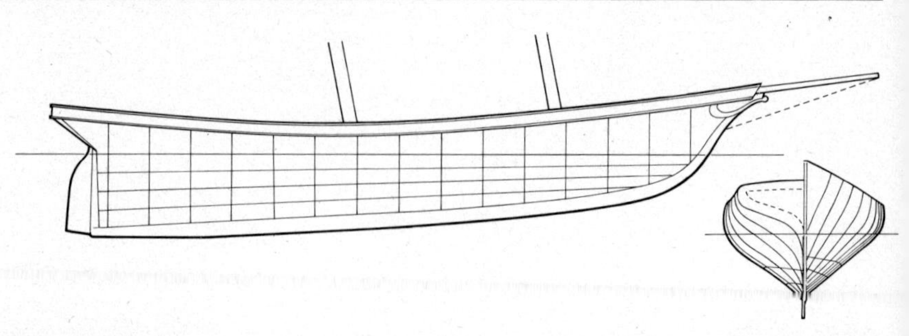

The hull shape of the schooner *America*, 1851.

Just 42 years after *America*'s victory a contemporary account said: 'The racing of the *America* in England has without doubt had more influence, directly and indirectly, on the yachting world than the performance of any other yacht, and both countries owe to her designer and owners a debt of gratitude that will remain uncancelled for generations; for it has been the means of bringing the two yachting nations together in many friendly contests, resulting not only in markéd modifications in the form and rig of the yachts of both countries, but the social intercourse begun so many years ago has continued and increased greatly to the benefit of yachting, and has led to a more complete union of all interested in the promotion of close international relationships.'

The first ever race for the Hundred Guinea Cup started at 10.00 on 22 August 1851. The yachts were anchored before the start at Cowes and cast off on the starting-signal. The British fleet of 17 yachts varied in size from *Aurora*, 47 tons, to *Brilliant*, a 392 ton three-masted schooner. The race was to the eastward: that is out to the Nab lightship, then to St Catherine's Point and back into the Solent via the Needles. The wind was light southwesterly but rather variable. At the Needles, *America* was eight miles ahead of the much smaller *Aurora*, but as she came back up the Solent against a foul tide and a failing wind, *Aurora* closed the gap. *America* crossed the finish off the RYS 10 hours 37 minutes after the start and *Aurora* crossed 10 hours 55 minutes – 18 minutes later. By any reasonable time allowance system, *Aurora* would have been the winner, but the race was not on time allowance. Anyway *America* had been well ahead until she hit the foul tide at the Needles, the British were in their own particularly tricky waters and the British yachts of *America*'s own size were a long way behind. However there is no basis at all for the most apocryphal remark in yachting history, where Queen Victoria is said to have been told of the victory of the American yacht and when asked who followed was told 'There is no second, ma'am.'

The Cup first became an international challenge trophy in 1857, when the surviving members of the *America* syndicate gave the cup (which they had won outright) to the NYYC. They stipulated that it be available for any foreign yacht club to challenge with a yacht of between 30 and 300 tons.

The first challenge for the Cup was by James Ashbury owner of the schooner *Cambria*. He was of humble origin, was not an experienced yachtsman and was not the first man to find that an America's Cup challenge was effective in making him of some note. His 108 ft two-masted yacht was the only challenger to have to take on the whole NYYC fleet simultaneously! Fourteen yachts defended in a single race. Time allowances were applied (as in every subsequent race until 1930). *Cambria* was 10th, finishing 27 minutes 3 seconds after the winner, the schooner *Magic*. *America* competed and finished 4th. *Cambria* was the only challenger to be subsequently used for commerce: she was sold to the West African coasting trade.

The second challenge for the America's Cup involved an arrangement unique in yacht racing. Although the challenging yacht, *Livonia*, a 127 ft schooner designed by Michael Ratsey and again owned by James Ashbury, only sailed against one defender, there were four yachts available to the NYYC. Depending on the weather of the day, the NYYC selected the most suitable defender. For this second challenge, raced in 1871, it was one of the few times that no defender was built. *Livonia* defeated one of the defenders, *Columbia*, in the third race: this was the first race ever won by a challenger and the last until 1920.

The year 1871 saw also the first sustained ill feeling between a challenger and the NYYC. It was the only time that a challenger claimed to have won the series. Ashbury claimed that he had protested in the second race and this should have been allowed; therefore he had won two races. The NYYC ignored this, but Ashbury subsequently sailed over the course and on a further day sailed over the line. So he judged he had won four races and the Cup! The bad feeling which resulted was one of the reasons why there was no further British challenge for 14 years.

There have only been two Canadian challenges for the Cup – in 1876 and 1881. The first Canadian challenger was a 107 ft schooner, *Countess of Dufferin*. It was the first series in which a single defender was nominated for all races. The challenger owned by Major Charles Gifford and syndicate was beaten in two races by *Madeleine*, which secured the Cup. The challenger was one of the worst equipped ever. Her sails were reported to fit 'like a suit of sailor's dungarees' and they were re-cut on arrival in the United States. The series was also unusual in that *America* followed one race round and put up a better time than the challenger. The second Canadian challenge in 1881 was by *Atalanta*, owned and sailed by Alexander Cuthbert of the Bay of Quinte Yacht Club, a small club that was keen to make a name for itself. The challenger, the first single-masted yacht to try for the Cup was soundly beaten by the 67 ft *Mischief*. In the second of her two victorious races, she won by a margin of 38 minutes 54 seconds which have never since been exceeded.

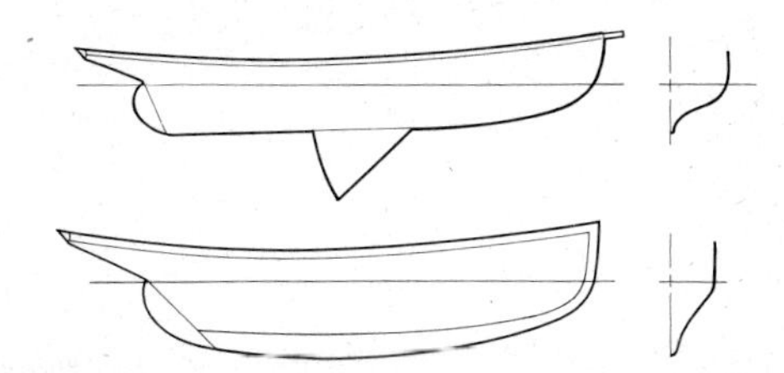

The remarkable contrast between (top) the American defender of 1885, *Puritan*, and (bottom) the British challenger *Genesta*. Each was designed to the rating rules and customs of its own country. The American was a beamy centreboarder, the British boat was narrow with a deep keel.

The most sportsmanlike gesture in a Cup series occurred in the next challenge by Sir Richard Sutton with the 96 ft cutter *Genesta*. In the second race (the first had been postponed due to lack of wind), the defender *Puritan* on the port tack collided with *Genesta* and should have been disqualified giving the race to the challenger. The committee said so but Sir Richard retired as well, saying 'We are very much obliged, but we don't want it that way. We came over for a race not a sail over.' *Puritan* went on to win the two necessary races, but by the closest margins so far. For this series and all future ones until after 1945, the challenging yacht was obliged to sail to the location of the races off New York 'on her own bottom'; so from England, it was necessary to have a yacht which could sail the Atlantic. *Puritan* was later sold to owners in the Cape Verde Islands and was used in the early twentieth century as a freight and passenger vessel between there and the United States.

The first challenge using a yacht of steel was by Lieutenant William Henn, RN in 1886 in his 103 ft cutter *Galatea*. Amazingly to modern ideas on yacht design and racing, she was not built to the rating rule under which the races were to be held. This was the American cubic contents rule, while *Galatea* was to the current English tonnage rule. The defender, *Mayflower*, designed by Edward Burgess and owned by General Charles J. Paine, was roughly the same length over all and on the waterline but had beam 23 ft 6 in, draught 9 ft 9 in (with centreboard to 20 ft) while *Galatea* was 15 ft beam, draught 13 ft 6 in and fixed a heavily ballasted keel. The boats were very unequal and *Galatea* was defeated in two straight matches, in one by the huge margin of 29 minutes 9 seconds.

The old 'inside course' near New York Harbour was used for the last time in the seventh challenge, which was in 1887. The Scottish challenger *Thistle*, sailing under the flag of the Royal Clyde Yacht Club, owned by James Bell and syndicate and designed by George L. Watson was defeated in two straight races by *Volunteer*, which had the same owner and designer as *Mayflower* and had held trials against her.

The most important alteration of the rules to be made until 1956 came into force in 1887 over the signature of the Commodore of the NYYC and also George Schuyler one of the original *America* syndicate in 1851. The main points were to specify that the yachts must be between 65 and 90 ft waterline; that races would be in open water away from headlands and shoals; that the challenging club must provide basic dimensions of its yacht ten months in advance of the series. The purpose of the last will be understood when it is remembered that the races were on time allowance and the challenger was measured on arrival at New York. Once they began to race on level rating (in 1930) the matter was no longer of significance. In practice the NYYC never enforced it and only asked for the length of the challenger. Another step towards modern yachting practice was making each Cup

Nathanael Herreshoff

series the best three out of five races rather than two out of three. From 1920 it became the present four out of seven.

The worst of any bitterness and bad sportsmanship to arise in the America's Cup was in the second of the two challenges by the Earl of Dunraven under the flag of the Royal Yacht Squadron. The feud continued up to a point where Lord Dunraven accompanied by his barrister appeared at a special inquiry at the New York Yacht Club, the club committee being enlarged by the famous Captain A. T. Mahan, USN and H. J. Phelps, a recent US Minister to Britain. Lord Dunraven's charges were quite unproven and, as he also failed to apologize, he was expelled from the NYYC, of which he had been an honorary member. The disputes received wide coverage in the world's Press, making the Cup races more of an institution than ever before.

Dunraven challenged with *Valkyrie II* in 1893. The defender was *Vigilant* selected after trials with three other newly built contenders. This process as so often before and since spelt defeat to the lone foreign yacht. All the boats were built to the length and sail area rule, so were more similar than in previous years and the competition was close. The victorious *Vigilant* was LOA 124 ft, waterline 86 ft, beam 26 ft 3 in, draught 14 ft, but draught 24 ft with the centreplate down. Her sail area was 11 272 ft^2. The era of the big cutters had begun in earnest. *Valkyrie II* had a little less beam and no centreplate as was the English tradition.

Vigilant was designed by Nathanael Herreshoff of Bristol, Rhode Island. One of the greatest yacht designers of all time, he designed the next five successive defenders of the Cup. The only designer to approach this is Olin Stephens who, up to 1974, had designed four boats and been involved in the design of two more. One, *Ranger*, was in conjunction with Starling Burgess and one, *Intrepid*, had been modified for her second Cup by Britton Chance.

Valkyrie II was defeated in three straight races, but in the third race only

Left to right
Lord Dunraven, whose name became a byword for disputes in yacht-racing.
Valkyrie II.
Vigilant, 124 ft.

Sir Thomas Lipton

by 40 seconds, the second closest result in Cup history (*Weatherly* beat *Gretel* by 26 seconds in 1962).

Lord Dunraven's complaints were mild in this first challenge of his; he complained of wind shifts and wash put up by spectator craft. *Valkyrie II* was sunk on the Clyde the following year in a notorious collision with the 90 ft cutter *Satanita* (see p. 246).

The arms race in cutters having begun under the length and sail area rule *Valkyrie III*, of 1895, was bigger: LOA 129 ft, sail area 13 028 ft^2. Herreshoff countered with *Defender*, slightly smaller but collecting 29 seconds allowance over the course. After the first race and a defeat by 8 minutes 49 seconds, Dunraven accused the defending yacht of having had ballast put aboard after measurement. He again complained about wash from spectator boats. Both competitors were as a result, remeasured, found to be in order and the second race was sailed. A few seconds before the starting-gun, a spectator craft cut between the two competing yachts, escaping herself but causing a collision between them. *Defender*'s rigging was partly damaged by *Valkyrie III*, but she finished the race, though losing to the challenger. However *Defender*'s subsequent protest was upheld and Dunraven's boat was disqualified. Two days later the third race began. Both boats crossed the starting-line, but *Valkyrie III*, turned away, hauled down her flag and returned to her anchorage. It was the end of the series, but not of the arguments. Back in England, Dunraven wrote a letter to *The Field* in which he raised again the whole question of ballast being altered to cheat the measurement. The results of the protest case also reverberated with Dunraven making it clear that the committee had arrived at the wrong decision. The sequence was the Earl's return to New York in December 1895 to appear before the NYYC inquiry.

The most persistent of all challengers was Sir Thomas Lipton. He restored the good feeling between British and American yachtsmen that had been tarnished by Lord Dunraven. He purposely made the Cup even more of an

Left to right
Shamrock, Sir Thomas Lipton's first challenger of 1899.

Columbia, only Cup defender to sail twice running in unaltered hull form. Here she lies ahead of *Shamrock*. Both boats are rigging the unwieldy spinnaker poles of those days.

The biggest single-masted yacht in terms of length and sail area, *Reliance*. The Cup defender of 1903, she was Herreshoff's defender. For her was coined the name 'racing machine'. After her the length and sail area rule under which she was designed was abandoned in favour of more controlled methods of rating.

Charlie Barr

event that was publicized in the Press of the world. A self-made millionaire in the tea business he used his Cup challenges to promote that business and enlarge his prospects in the American market. But always he took a sporting and generous attitude, right through his five defeats between 1899 and 1930. Born in Ireland and raised in Scotland, he always sailed under the flag of the Royal Ulster Yacht Club. It was only after his last challenge and a year before he died at the age of 81 that the Royal Yacht Squadron elected him to membership; before that his election had been blackballed on various grounds. Some people would, however, understand the Squadron's dislike for his initial motives, though its recognition of his ultimate sportsmanship. But the events show the lack of uniform support among British yachtsmen for America's Cup enterprises.

The first Lipton challenge, in 1899 was by *Shamrock*. She was towed across the Atlantic by Lipton's steam yacht *Erin*, the first time that a challenger had been brought over in this way. It became standard method until, in 1956, the challenger could be shipped. *Shamrock*, designed by William Fife, was beaten in three straight matches by *Columbia*, Herreshoff's third defender. She was owned by a syndicate headed by J. P. Morgan the famous tycoon and banker. Both boats were close in size under the length and sail area rule. The defender had an LOA of 131 ft, LWL 90 ft, sail area 13 135 ft^2.

The second Lipton challenge was with *Shamrock II* in 1901. She was designed by G. L. Watson. She was the longest ever challenger at LOA 137 ft. Once again she lost in three straight matches to *Columbia*, the only defender to race twice under the same skipper Charlie Barr and in unaltered form. (*Intrepid* defended in 1967 and 1970 but was greatly altered and not by her original designer.) However the defence was preceded by great rivalry: a defence candidate from Boston owned by Thomas W. Lawson was built. The New York Yacht Club insisted that a member of the club defended the Cup and invited Lawson to join, but he refused in order to gain honour for Boston. The NYYC did not invite him to their trials, but

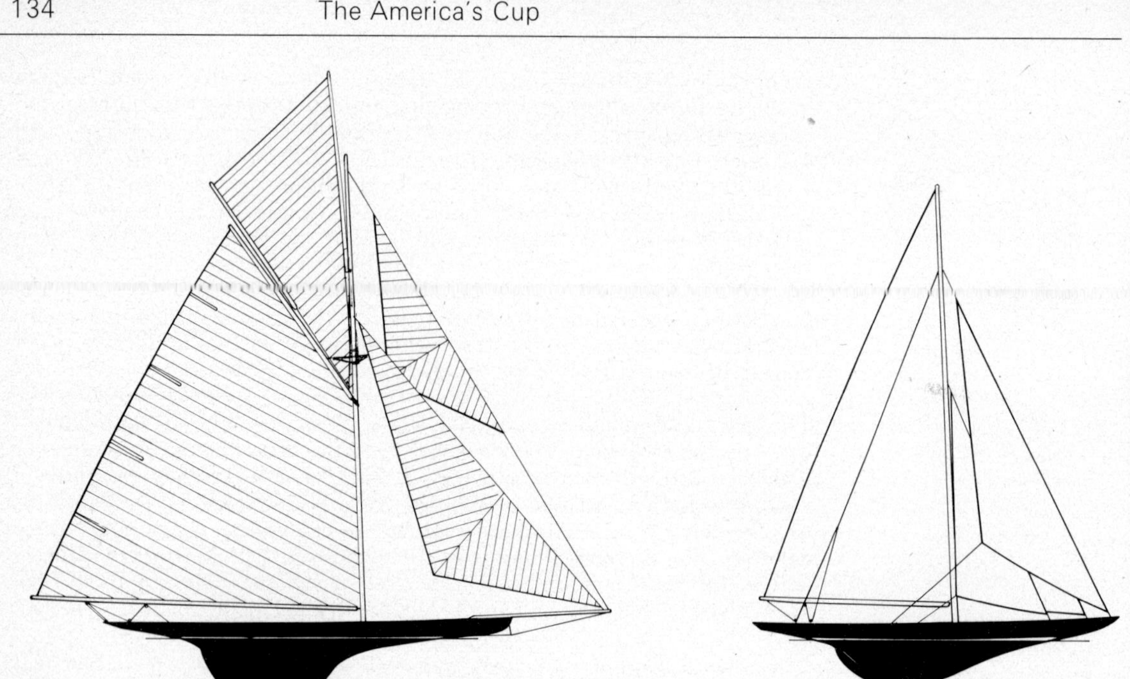

Left to right
A comparison of the largest ever America's Cup yacht, *Reliance*, 1903, a J-class yacht of the 1930s, and smaller still, a 12-metre in which the cup is sailed in the 1970s.

an independent club at Newport, R.I. did so and Lawson's boat, *Independence*, came last. *Shamrock II* was the only challenger never to return to her home country. She was broken up in New York.

The largest and most extreme yacht ever to sail in an America's Cup match, indeed any yacht race, was the defender of the 1903 series, *Reliance*. It was no coincidence that this was the last match before the First World War and huge nearly untaxed fortunes could be amassed in America and Britain as never before or since. *Reliance* was the ultimate development under the length and sail area rule – a huge skimming dish. To her was first given the nowadays oft-used expression 'racing machine'. Herreshoff's fifth defender, she was owned by C. Oliver Iselin and syndicate and measured LOA 142 ft 8 in, LWL 89 ft 8 in, beam 25 ft 8 in, draught 20 ft. There was a lead keel and she was built in bronze. The biggest sail area ever put on a single-masted yacht gave *Reliance* 16 160 ft^2. There was a steel mast and wooden topmast for the gaff rig and its topsail. Her skipper was Charlie Barr, the little Scots American and only man to be professional skipper and helmsman of three Cup defenders. The crew were mainly Scandinavian. *Reliance* was chosen to defend after trials against *Columbia* and *Constitution*.

In this 1903 series, for the first time a race was postponed owing to there being too much wind. This was a reflection on the extreme type of yacht the length and sail area rule had allowed. After this the Universal Rule, which demanded displacement, was used for the Cup races. *Shamrock III*, Sir Thomas's third challenger designed by William Fife lost in three races. In the last race she failed to find the mark in the fog and did not bother to cross the line. Yet another example was given of the many times on which a challenging yacht proved herself far inferior to the defender.

The only Englishman ever to threaten American dominance in the design field was Charles E. Nicholson. He designed Lipton's fourth challenger, *Shamrock IV*. She was the most successful of the *Shamrocks* and the first challenger to the Universal Rule. This *Shamrock* was bigger than the defender *Resolute* at LOA 110 ft 4 in; LWL 75 ft; 10 458 ft^2 sail area. Three American boats were built from which *Resolute* was selected: the others were *Defiance* and a Boston boat *Vanitie*. The First World War broke out when *Shamrock* was on her way across the Atlantic in August 1914. She was laid up in New York until 1920.

The 1920 races were the last to be held off the outer course of New York Harbour near Sandy Hook. They were the last to be held using time allowance. They were the first to have in the event as many as five races to decide the winner. It was the first time that both boats had amateur skippers – another sign of the changing mode of yacht racing. The American was Charles Francis Adams and the Englishman was William Burton. The first two races were won by *Shamrock IV*. This was not only the first win by a challenger since 1871 but the first time that a British yacht had won two races in succession let alone at the beginning of a series – it looked as though there was a real threat to the NYYC record.

In the first race *Resolute* had been disabled because her throat halyard jumped the block and the goose neck of the gaff jumped the mast so she did not finish and the challenger took the race. In the next race in light weather, *Shamrock*'s lead was 10 minutes 5 seconds on elapsed time and 2 minutes 26 seconds on corrected time. She rated much higher owing to her great sail area. It only required one more race by the challenger to take the Cup back to Britain. But *Resolute* won the next three races by 7 minutes 1 second (the elapsed time was an exact tie – the only dead heat in Cup history), 9 minutes 58 seconds and 19 minutes 45 seconds.

The first America's Cup series to be held on the present course off Newport, R.I. was in 1930 after a gap of ten years when Lipton, by now an old man of 80, challenged again. He was loath to leave New York and the publicity it afforded, but the commercial traffic made it essential to seek better racing waters. The yachts were now to the Universal Rule's J class of 76 ft rating. From now on the boats raced level: no time allowances were applied. No less than four potential defenders were built in the United States for the honour of competing with *Shamrock V*. They were *Enterprise*, *Weetamoe*, *Yankee* and *Whirlwind*. It was the old and to be repeated story of a single

Shamrock III, at her launching. Designed by William Fife, she was hopelessly outclassed by the defender *Reliance*; a situation which has occurred too frequently in America's Cup matches.

The only man to skipper and steer three successive Cup defenders, Harold S. Vanderbilt.

challenger against a competitively selected defender. During the trials *Yankee* put up a then record for the 30-mile America's Cup course of 2 hours 47 minutes 59 seconds. As all the boats rated at or just below 76 ft they did not vary greatly in size. An example was the actual defender *Enterprise* with LOA 120 ft 11 in; LWL 80 ft; beam 22 ft 2 in; draught 14 ft 7 in; displacement 128 tons and sail area 7583 ft². For the first time and ever since, each yacht had a Bermudan mainsail. The foretriangles had three small sails – jib topsail, jib and forestaysail. Genoa jibs also new were only used in very light winds.

The only man to skipper and steer three successive and different Cup defenders was Harold S. 'Mike' Vanderbilt, one of the greatest yachtsmen of all time. He skippered *Enterprise* which had been designed by W. Starling Burgess. The yacht had many new features including a light alloy mast and a 'Park Avenue' boom with a wide section on which the huge mainsail took up its required shape. *Shamrock V* was designed by Charles E. Nicholson and skippered by Ernest Heard. She was beaten in four straight matches, in one of which she retired when the main halyard parted. It was one of those occasions to be repeated in 1937 and 1967 when American technology and sailing ability leapt ahead and quite outclassed the challenger. It was indeed the most decisive defeat of Sir Thomas Lipton. At the end he said 'I willna challenge again. I canna win.' The famous remark was typical of the thoughts at some time of every challenger.

The challenge that came closest to winning the America's Cup was that of the flying ace and later millionaire aircraft-manufacturer, T. O. M. Sopwith, in 1934. His J class yacht *Endeavour*, designed by Charles Nicholson was faster than the defender and he won the first two races and lost the last by less than a minute. He failed to sustain his early success because he was outwitted by the American helmsmen and had an inferior crew. This latter problem was owing to a strike for higher pay by his professional crew before leaving England. As a result he had his Captain, Williams, nine faithful professionals and a balance of amateurs. *Endeavour* carried the first quadrilateral jib, though this was spotted in trials and copied by the Americans. By this time also spinnakers had become symmetrical, instead of the earlier triangular sort sheeted to windward of the forestay. *Endeavour*'s dimensions were LOA 130 ft; LWL 83 ft 4 in; sail area 7561 ft². The defender *Rainbow* cost $500 000, was owned by Mike Vanderbilt and syndicate and

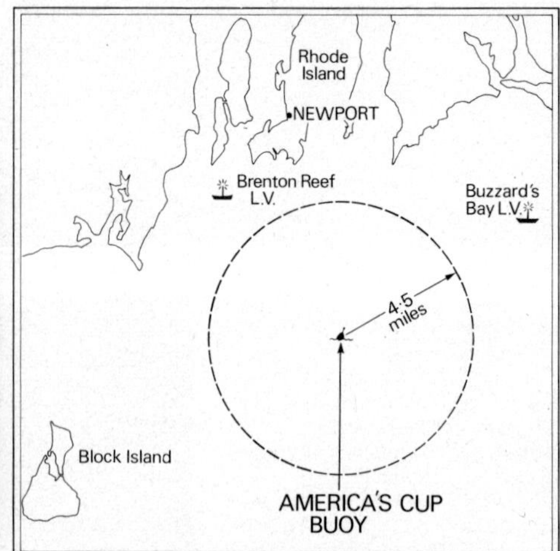

Shamrock V, Lipton's last challenger, beaten in four straight races in 1930. It was the first British challenger to the American J-class (76 ft rating) rule.

Right
Endeavour, the best of any British challenger, looked set to take the Cup, but America's yachtsmen rallied just in time.

steered by Vanderbilt. She was designed by Starling Burgess. She beat the trialists *Yankee*, *Vanitie* and *Weetamoe*: Olin Stephens was crewing in the last.

The circumstances of this cliff-hanger of a challenge were that *Endeavour* won the first race by over 2 minutes and the second by 51 seconds. In the third race she was leading by over 6 minutes at the leeward mark, but the wind went light which British yachtsmen traditionally dislike. At this time Vanderbilt turned the wheel of *Rainbow* over to Sherman Hoyt: by clever tactics among which Sopwith ran into light air, Hoyt won the third race. What happened on the evening of that race demonstrates the sort of action the defenders will take if they think there is a chance of losing. Despite the ancient rivalry between New York and Boston, Frank C. Paine, owner of the Boston boat *Yankee*, was asked to join *Rainbow*. He did so with his own J class spinnaker – the finest in America. The ballasting of *Rainbow* was altered – so they changed both the men and the ship.

Left
Shamrock V, seen in 1975 after returning to her builders Camper and Nicholsons for refitting. Since her racing days she had been in Italian ownership as a cruising yacht.

After 1920, the America's Cup races were moved from New York, which had become too congested, to the waters of Newport, Rhode Island. The yachts start from the 'America's Cup Buoy' and the course is now laid out Olympic fashion, with a 4·5 mile leg to windward.

The most famous of all protest cases was the feature of the fourth race of the 1934 America's Cup. On the reaching leg *Endeavour* luffed *Rainbow*. *Rainbow* with Vanderbilt at the helm kept going and Sopwith bore away to avoid the serious collision that would result between two J boats. *Rainbow* kept ahead for the rest of the race and *Endeavour* crossed the line 1 minute 15 seconds later with a protest flag in the rigging. The next day the NYYC committee rejected the protest that *Rainbow* had failed to respond to a luff on the grounds that the protest flag had not been hoisted immediately after the incident. There was anger then and for a long time after. The third of the immortal quotes of the Cup races was coined – all three concern *defeat* – 'Britannia rules the waves, but America waives the rules.'

Rainbow won the next two races. In one of them brilliant spinnaker tactics by Paine were largely responsible and in the final one *Endeavour* lost by 55 seconds. The hands of many of the amateur crew of the challenger were in bad shape by this time which had limited Sopwith in his choice of

Ranger, the 1937 defender and last of the J-class

Right
The sight that was seldom seen and will certainly never be seen again. Three American and two British J-class yachts, each of them over 130 ft long race as a class after the 1937 America's Cup matches.

quick tacking and similar tactics, though their observed performance was professional in handling and said to be no worse than *Shamrock V*. The greatest chance that the Cup ever had of returning to the Royal Yacht Squadron (under whose flag Sopwith had challenged) was lost.

The last racing of really big yachts inshore took place in the 1937 series. It was the last of the J class and saw the ultimate yacht of the class, the defender *Ranger*. She was owned by Mike Vanderbilt. Even in 1937 such yachts were too expensive and no syndicate was found to support her, nor were there any other new contenders for the defence: *Rainbow* and *Yankee* took part in the trials. *Ranger* was vastly superior, designed by Starling Burgess and with considerable assistance from the young Olin J. Stephens, whom with his brother Rod also sailed in the yacht. Charles E. Nicholson had presented *Endeavour*'s lines to Burgess after 1934 and these were put in the tank with other models. The chosen model among those tested by Burgess under the orders of Vanderbilt is thought to have been one of those to the lines of Stephens. She was LOA 135 ft 3 in; LWL 87 ft; beam 21 ft; draught 15 ft; sail area 7546 ft². Her mast was 165 ft high. The amount of sail worked into the foretriangle was greater than ever for the measured area. The spinnaker was the largest sail ever made at 18 000 ft² (more than two-fifths of an acre). The hull was of steel. During trials *Ranger* sailed a new record over the 30-mile course of 2 hours 43 minutes 43 seconds (11·1 knots). *Endeavour II* had been designed by Nicholson without the tank but using traditional yacht-design methods. She was faster than *Endeavour* and her professional crew were smarter than the amateurs of 1934. But *Ranger* was altogether faster than her rival and won four straight races, two of the margins being large: 17 minutes, 18½ minutes, 4½ minutes and 3½ minutes.

In the third race *Ranger* achieved the all-time record for a yacht on a windward leg: 15 miles to windward in 2 hours 3 minutes 45 seconds.

After the Cup, the J class raced in American waters. Never again were such yachts to be seen anywhere in the world in such a group or indeed racing at all. In the 13 races, *Ranger* won 12, *Endeavour II* had seven second prizes, *Rainbow* four and *Yankee* one. *Endeavour* sailed the races as well. The American

Vim (left) and *Columbia* in the 1958 trials when 12-metres started to be used for the America's Cup. *Vim*, a twenty-year-old Stephens' design, became the starting-point of American 12-metre superiority. *Columbia* just beat her in trials and became the successful defender.

Right
1962, the first-ever Australian challenger, *Gretel*, beats the American defender *Weatherly* in two races. Here she surges over the finishing-line to win the second race.

J boats were broken up during the Second World War for scrap. Both *Endeavours* lay mastless, but with other spars in the mud on the Hamble River until 1970. To campaign such a yacht in the money values of 1975 would cost of the order of £2 000 000 even if crews could be found to handle such boats. But by 1945 they were an anachronism. In 1972 *Endeavour* was surveyed with a view to being rigged for crossing the Atlantic for renovation in the USA, but was found quite unfit for this. The Maritime Trust, London, decided to restore her as a static monument of the time of the big racing cutters but, at the time of writing, she lies at Cowes awaiting a sponsor for this project. In 1975, *Shamrock V* came to England for a refit, being in use as a cruising yacht in Italian ownership.

Late in 1937 Uffa Fox wrote: 'We on this side have learnt some lessons and let us hope a new challenger will soon be on her way across the Atlantic with a firm endeavour to bring back the Cup once more to these shores.' But far from being 'soon' the longest period during which an America's Cup was not sailed was from 1937 to 1958. The reasons are obvious with the social and financial changes brought about by the Second World War. In 1956 the US Supreme Court allowed changes to the 1887 Deed of Gift of the Cup to reduce the minimum waterline from 65 to 44 ft and to require no longer that the challenger cross the Atlantic on her own bottom. This enabled the 12-metre class yachts to the International Rule to be used for the races (see p. 193). They had already been racing regularly as a class in both America and England before 1939. Apart from being just about half the length of the J class, they had to be built of wood and could be manned by an amateur crew. There were some rules for accommodation but the yachts came to be designed entirely for inshore racing and by 1970 the accommodation rules were officially dropped.

Another real chance was thrown away by the British with the first post-war challenge. The 12-metre class had been sailed far more in England than in the USA and now also the disadvantage of sailing the Atlantic had been removed. But the old, old mistake of building just one challenger was made. The designer was David Boyd of Scotland, who was not in the business of designing yachts to modern competitive rating rules, but had designed the

American contenders fight for the honour of defending yacht in 1964. *Constellation* was chosen, but any of the American boats could easily have beaten British challenger *Sovereign*.

winning 6-metre *Lalage* before the war. The new *Sceptre* had a clever innovation of a huge cockpit in the centre of the boat – later the rule was changed to stop such practice. She was built on the Clyde and had trials in the Solent against Owen Aisher's pre-war 12-metre, *Evaine*. The writing was on the wall when *Sceptre* found she had difficulty in consistently beating *Evaine*.

In the USA the starting point was another pre-war 12-metre designed by Sparkman and Stephens, where Olin Stephens was in charge of design. Altogether for 1958 there were built three new 12s: *Columbia*, *Weatherly* and *Easterner*. *Vim* sailed by Emil 'Bus' Mosbacher started by doing better than the others. But *Columbia* was also a Stephens design sailed by Briggs Cunningham, but with Olin and Rod Stephens aboard as they had been on *Ranger*. She won the trials. She went on to beat *Sceptre*, sailed by Graham Mann under the flag of the Royal Yacht Squadron (a syndicate owned her) very soundly. The margins on courses which were now 24 miles were 7 minutes 44 seconds, 11 minutes 42 seconds, 8 minutes 20 seconds, 6 minutes 52 seconds.

The first challenge for the America's Cup by Australia was in 1962 and also the closest series since *Endeavour* in 1934. The Americans did not rate the Australian effort highly in advance and only one new yacht, *Nefertiti*, was built, to the design of Ted Hood. The new boat with *Columbia*, *Easterner* and *Weatherly* took part in the trials. *Weatherly*, designed by Philip L. Rhodes and sailed by Bus Mosbacher and owned by Walter Gubelmann and syndicate, was selected as defender.

Gretel owned by Sir Frank Packer, designed by Alan Payne and sailing under the flag of the Royal Sydney Yacht Squadron was, like *Endeavour*, the faster boat. She won the second race and lost the others by the small margins of 47 seconds, 8 minutes 40 seconds, 26 seconds and 3 minutes 40 seconds. It is still the closest series seen since the war, closer than all

Intrepid in 1970. She became the second yacht to defend the America's Cup twice, having already beaten the Australian challenger in 1967. She only just missed being chosen as defender again in 1974 and so becomes one of the classic yacht designs of all time.

subsequent ones. The 26 second margin is the closest-ever result in a Cup race, where corrected time was not involved. Mosbacher saved the Cup by superior helmsmanship and tactics. As a result he became the first Jew to be elected a member of the New York Yacht Club.

The next challenge was by Britain and resulted in the worst margin of defeat since *Mayflower* beat *Galatea* in 1886 – and it will be remembered that the old boats were not even built to the same rule. In 1964, it was the 19th America's Cup challenge and Britain's last (to date). For the first time the challenging country held its elimination trials off Newport, R.I., which should in theory have greatly increased the chances. However the trials were between two boats of similar design, both by David Boyd. *Sovereign* was owned by Anthony Boyden and *Kurrewa* was owned by Australians Frank and John Livingston and run under the control of Owen Aisher. *Sovereign* won the trials, but the writing was again on the wall because neither boat had seemed vastly superior to *Sceptre*.

The greatest ever number of American contenders for the defence took part in the 1964 trials. New boats were *American Eagle* and *Constellation* and they were joined by *Columbia*, *Nefertiti* and *Easterner*. *Constellation*, designed by Sparkman and Stephens and skippered by Bob Bavier was chosen. Peter Scott was at the helm of *Sovereign*. The margins in four straight races were 5 minutes 34 seconds, 20 minutes 24 seconds, 6 minutes 33 seconds and 15 minutes 40 seconds. After this the NYYC felt its members had built new yachts for nothing and British yachting prestige was at a low ebb. The club ruled that challenges would not be accepted more frequently than every three years. The subsequent matches have been in 1967, 1970 and 1974 and all from Australia.

The 1967 matches at last saw the challenging nation, Australia, with two contenders: the new *Dame Pattie* owned by Emil Christiansen and syndicate and *Gretel*, much modified. *Dame Pattie*, named after the Australian Prime

Minister's wife, designed by Warwick Hood and skippered by Jock Sturrock, was the challenger. One new American boat was built: *Intrepid*, designed by Sparkman and Stephens and steered by Bill Ficker. The design was a considerable advance over previous 12-metres rather as *Reliance* and *Ranger* had been in their day. She won the trials over *American Eagle*, *Columbia* and *Constellation* and went on to defeat *Dame Pattie* in four straight races. *Intrepid* brought a new shape to 12-metres with a bustle, a sort of swollen area in the after part of the yacht under water and a separated rudder on the stern. Such configurations were at the same time coming into universal favour for ocean racing yachts.

In 1970, *Intrepid* became the only yacht besides *Columbia* in 1899 and 1901 to defend the Cup twice running. However she was considerably modified by Britton Chance who was not, of course, her original designer. In 1974 she was nearly selected as the final trials were between her and the eventual defender *Courageous*. However the trials were so close that she could also have beaten the challenger. That time she had been modified yet again, but by her original designer Olin Stephens.

The first ever French challenger, *France*, owned by Baron Bich went to Newport, R.I. in 1970. She was eliminated in four straight races by the Australian challenger, *Gretel II*, owned by Sir Frank Packer designed by Alan Payne and skippered by Jim Hardy. In the American trials the first-ever contender from the southern United States took part, *Heritage*, designed and built in Florida by Charles Morgan. Another new yacht was the Sparkman and Stephens *Valiant*; the old *Weatherly* also competed – all in vain against *Intrepid*.

The series was notable for the greatest acrimony since Lord Dunraven accused his opponents in 1895; the first collision since Sir Richard Sutton in 1885; and the first protest since 1934. In the first race Hardy protested over an alleged infringement against *Gretel II* before the starting-gun but the protest was dismissed by the NYYC. It was in the third race that the collision occurred just after the starting-gun, since the boats had touched there had to be protest flags flown by both sides. The Australian yacht finished first. The fat was then in the fire because in the resulting protest hearing the NYYC committee dismissed the Australian protest on perfectly legal grounds, but this meant the Australian win was taken from them and *Intrepid* won.

The resulting publicity, much of it ill informed, was mainly directed at the NYYC committee. One Australian Member of Parliament said that the Ambassador should be recalled from Washington and the Australian troops withdrawn from serving with the Americans in Vietnam! One result was that in 1974, there was an international jury with 'neutral' members.

The third race was won by *Intrepid*, but the fourth race was won by *Gretel II* by 1 minute 2 seconds. This was the first time a challenger had ever won a race just losing a sequence of three. However the series ended when Bill Ficker, the first Californian – and west coast yachtsman – to skipper a defender, beat Hardy by 1 minute 44 seconds.

France, the French challenger for the America's Cup. She was eliminated in the preliminary trials in 1970 and again in 1974.

The most overtly commercial challenge was that by the Australian Alan Bond, a property millionaire from Western Australia and, under 40, the youngest owner to challenge. He challenged under the flag of the Royal Perth Yacht Club but the challenge was meant to enhance the prestige and value of Yanchep Sun City, a huge holiday development. His dream was to hold the next series there. But once again the mistake was made of having a single challenger, *Southern Cross*, and she was designed by a comparatively

RECORD OF MATCHES

Winning American Yacht				Challenging Yacht			Series best of so many races	Number of races	Average course length
Time allowance by:	**Year**	**Name**	**Skipper**	**Name**	**Country**	**Skipper**			
Waterline area	1870	*Magic*	A. Comstock	*Cambria*	England	J. Linnock	1	1	35·1
Displacement rule	1871	*Columbia*	N. Comstock	*Livonia*	England	J. R. Woods	5	3	36·6
Displacement rule		*Sappho*	S. Greenwood	*Livonia*	England	J. R. Woods		2	37·5
Cubical contents rule	1876	*Madeleine*	J. Williams	*Countess of Dufferin*	Canada	J. Ellsworth	3	2	36·3
Cubic contents rule	1881	*Mischief*	N. Clock	*Atalanta*	Canada	A. Cuthbert	3	2	32·3
Length and sail area	1885	*Puritan*	A. Crocker	*Genesta*	England	J. Carter	3	2	36·3
Length and sail area	1886	*Mayflower*	M. B. V. Stone	*Galatea*	England	D. Bradford	3	2	36·3
Length and sail area	1887	*Volunteer*	H. C. Haff	*Thistle*	Scotland	J. Barr	3	2	36·3
Length and sail area	1893	*Vigilant*	W. Hansen	*Valkyrie II*	England	W. Cranfield	5	3	30·0
Length and sail area	1895	*Defender*	H. C. Haff	*Valkyrie III*	England	W. Cranfield	5	3	30
Length and sail area	1899	*Columbia*	C. Barr	*Shamrock*	England	A. Hogarth	5	3	30
Length and sail area	1901	*Columbia*	C. Barr	*Shamrock II*	England	E. A. Sycamore	5	3	30
Length and sail area	1903	*Reliance*	C. Barr	*Shamrock III*	England	R. Ringe	5	3	30
Universal rule	1920	*Resolute*	C. F. Adams	*Shamrock IV*	England	W. T. Burton	5	5	30
Fixed rating by:									
Universal rule 76 ft (J class)	1930	*Enterprise*	H. S. Vanderbilt	*Shamrock V*	England	T. Heard	7	4	30
Universal rule 76 ft (J class)	1934	*Rainbow*	H. S. Vanderbilt	*Endeavour*	England	T. O. M. Sopwith	7	6	30
Universal rule 76 ft (J class)	1937	*Ranger*	H. S. Vanderbilt	*Endeavour II*	England	T. O. M. Sopwith	7	4	30
International rule 12-metres	1958	*Columbia*	B. S. Cunningham	*Sceptre*	England	G. Mann	7	5	24
International rule 12-metres	1962	*Weatherly*	E. Mosbacher, Jr	*Gretel*	Australia	J. Sturrock	7	5	24
International rule 12-metres	1964	*Constellation*	R. Bavier, Jr	*Sovereign*	England	P. Scott	7	4	24·3
International rule 12-metres	1967	*Intrepid*	E. Mosbacher, Jr	*Dame Pattie*	Australia	J. Sturrock	7	4	24·3
International rule 12-metres	1970	*Intrepid*	W. Ficker	*Gretel II*	Australia	J. Hardy	7	5	24·3
International rule 12-metres	1974	*Courageous*	T. Hood	*Southern Cross*	Australia	J. Hardy	7	4	24·3

Ted Hood

unknown designer, Bob Miller. She tuned up against *Gretel II*, but there were no eliminations in the American way. The 1974 challenge was the 22nd America's Cup series and the 4th Australian one.

Intrepid nearly became the only yacht to defend three times. Two of the American contenders *Mariner* and *Valiant* were eliminated early as their hulls were badly designed after the most mistaken deductions ever carried forward from a test tank into Cup trials. *Intrepid* was the first contender to be funded by public subscription and she came from the west coast. The racing between her and the new Sparkman and Stephens design was extremely close and there was nothing to choose between the boats. This was all the more remarkable because 1974 was the first time that aluminium hulls were allowed in 12-metres and the new boats were of this material, but not *Intrepid*. *Courageous* skippered by Ted Hood was selected by the NYYC.

Before these final knife-edge American trials, *Southern Cross* skippered by Jim Hardy had easily beaten the old *France*, which had appeared for elimination trials. The trials were run by the Royal Thames Yacht Club of London, the first time a foreign club had conducted eliminations for challengers in this way. As well as the international jury headed by Dr Beppe Croce, President of the International Yacht Racing Union, there was an international measurement team headed by James McGruer of Scotland. Thus the NYYC was moving over to the general practices of international regattas.

Courageous beat *Southern Cross* in four straight races and another race was abandoned because it did not finish within the time-limit. The margins were 4 mintes 54 seconds, 1 minute 11 seconds, 5 minutes 27 seconds and finally 7 minutes 19 seconds, thus leaving the total number of races ever won by a challenging yacht at 7. Since the war the only two were one each by *Gretel* and *Gretel II*. The same mistakes by the challenger and attitudes on both sides kept the NYYC firmly with the Cup in that impressive small room in the club. A single challenger against a selected defender after eliminations; a challenging owner with motives of commerce or fame as well as sport; an owner, too, of comparative recent entry into the world of yachting against the long-standing traditions of the NYYC and its members whose fathers and ancestors have upheld the American tradition of speed under sail; the immense resources in sailing men and designers possessed by America, now particularly Olin Stephens, unapproached in experience.

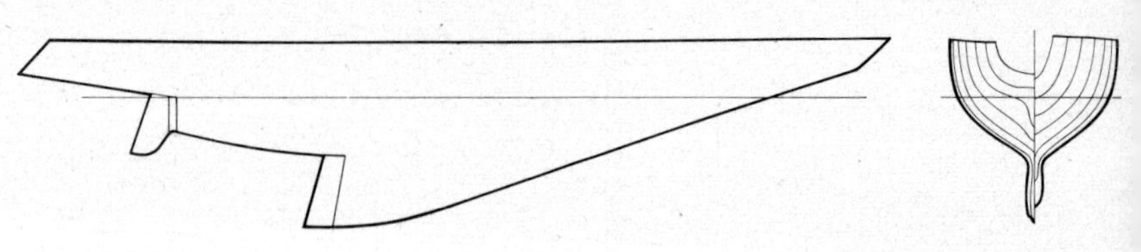

The hull profile and sections of the 1974 America's Cup defender, *Courageous*. The 12-metre rule prevents a complete fin, but the after part is cut away and ends are snubbed. LOA 66 ft 6 in, displacement 60 000 lb.

The 12-metres in which races have been sailed since 1958 are today the largest class of inshore racing yacht. Cost, of course, rises all the time, but the approximate cost of *Courageous* in 1974 to build, maintain and campaign her was $1 200 000. Her dimensions are little different from other 12-metres, though there is some choice of parameters under the International 12-metre Rule. The yachts have become rather shorter overall as modern ends are snubbed to save weight and gain rule advantage (*Gretel* and *Columbia* were 70 ft). *Courageous* was LOA 66 ft 6 in; LWL 45 ft 6 in; beam 12 ft; draught 8 ft 10 in; sail area 1700 ft^2; displacement 60 000 lb.

4 Trans-Ocean Sailing

CRUISING

'Cruising' is a term when used in relation to small vessels that is very much over used. It can mean any movement of the craft, which is not racing or the performance of some specific manœuvre. There are more descriptive words used when the yacht is cruising such as 'day-sailing', whereas 'cruising' might imply something more like a tour. A short run down the coast, taking 12 to 24 hours is sometimes called an 'overnight passage'. Sailing around for a few weeks might be for some people 'a holiday cruise'. Successive overnight passages and day-sailing from one place to another could be called a 'passage from X to Y'. 'Voyage' implies a round trip, be it short or long, unlike 'passage' which means a trip in one direction only. 'Voyage'

Typical modern cruiser. This American design, the Pearson 35, which has a centreboard is typical for week-end sailing.

Typical of a great mass of cruising boats, the small gaff sloop returns to harbour at dusk

would hardly be applicable to a short cruise: perhaps even less than about 500 miles would be an exaggerated use of the term. Sailing more than about 1000 miles without putting into a port, sailing for such a distance out of sight of land – the exception might be an isolated cape or island – or between land masses, comes under the heading of 'trans-ocean sailing'. Other terms used might be 'deep water cruising' – or 'passage making' or 'voyaging' – then reference might be made to 'long-distance' or 'blue water' cruising.

'Cruising' is a word used with several connotations, but it is maritime in origin. The Dutch word *kruisen* meant 'to cross' and a patrolling warship under sail would cross back and forth on patrol. Indeed only a man-of-war would have been inclined to cruise; merchantmen in the days of the origin of the term would sail as quickly as possible on their way to flee storms and enemies. Today cruising implies leisure whether in big cruise ships, or small craft, as used here. The naval meaning, of course, is still retained.

Cruising clubs

The achievement of a passage of 1000 miles in open water in a vessel under 70 ft is recognized by the Ocean Cruising Club. This is an association, with no actual club house, but whose Secretary is in England. Anybody who qualifies by sailing the distance in the appropriate craft, sail, power or rowing, is eligible for membership. The Ocean Cruising Club was founded in 1954 and by 1974 had 877 members, which is an indication of the extent of long-distance sailing. About 50 members join each year. About 40 per cent of the membership is British, 35 per cent is American and 25 per cent is of other nationalities, of which Australians form a sizeable part. The club has flag officers in Britain and specifically on the east and west coasts of the USA.

The two most distinguished clubs which have yacht cruising as their objectives are the Royal Cruising Club, founded in London in 1880 and the Cruising Club of America, whose secretariat is in New York. Neither have club houses, their membership being scattered round the world though chiefly in Britain and the USA respectively. They are not specially orientated to making long voyages or breaking records, indeed their senior members might well frown on such intentions. But in practice many of their members have achieved long passages. The Royal Cruising Club, which was founded by Sir Arthur Underhill, publishes annually accounts of some of its members' cruises, long and short. It has nothing to do with racing. The CCA has embraced several racing events, which were originally essentially for cruising yachts. These include the Newport, R.I. to Bermuda race, a leading ocean racing course. It was after the RCC elected not to have anything to do with the origin of the Fastnet race, that, in England, the Royal Ocean Racing was formed. In noting the existence of the RCC and the CCA, it must be remembered that most (though by no means all) yacht and sailing clubs are formed for the purpose of racing. Among the exceptions are the London-based (with club houses) Cruising Association and the Little Ship Club.

Early cruising

The first yacht voyages upon the ocean were in fact by what might be described as 'private ships'. One yacht with a vast distance to her credit was *Sunbeam* owned by Lord Brassey, Member of Parliament and a member of the Royal Yacht Squadron. Designed by Mr St Clare Byrne the *Sunbeam*

Lord Brassey's *Sunbeam*, 170 ft, as rigged for her voyage round the world.

had LOA 170 ft, with a displacement of 531 tons. She was a three-masted topsail-yard schooner with wooden on steel frames and with a single-screw-driven 350 hp engine. The bunkers contained 80 tons of coal, of which 4 tons were consumed daily if under power, She could carry 11 amateurs, which included the owner, his family, guests and doctor and 32 crew, including the lady's maid! Between July 1876 and May 1877 *Sunbeam* sailed a total of 37000 miles on a voyage round the world. The route was from Cowes, England into the South Atlantic, through the Magellan Strait, the Pacific, to Japan, the Indian Ocean, Red Sea and Suez Canal and through the Mediterranean back to the Channel – with many ports of call. In later years Mr Gladstone (in 1885) and Lord Tennyson (in 1889) were sailing guests of Lord Brassey.

The intense enthusiasm for cruising in the nineteenth century, by those who could afford it, is shown by Lord Brassey, who owned more yachts and covered more mileage afloat than any person known to have done so at that time. Between 1854 and 1893 he owned 14 different yachts and covered 228682 nautical miles. The yachts were not all large: from 1854 to 1858, he sailed *Spray of the Ocean*, an 8 ton cutter.

More common amateur sailing was that indulged in by amateurs who cruised on the 85 ft schooner *Hornet*, which had been built at Cowes in the middle of the century. She was manned by 35 enthusiastic amateurs and several professionals. Between 1879 and 1881, she was sailed to Ireland, Norway, Spain and Morocco.

The first regular cruising in small sailing-yachts was begun by Richard Tyrrell McMullen in the sloop *Leo* of only 3 tons and LOA 20 ft in 1850. Usually he sailed accompanied by a paid hand. It should be remembered that this was the period when yachting was emerging as a pursuit. The reasons, as have been mentioned elsewhere, were that after 1815 the seas were at last safe from marauders – before that no one would dream of going away from harbours and estuaries for pleasure – and that increasing affluence for some men allowed time for sailing for its own sake. Cruising in small yachts round the coasts of Britain was one aspect of yachting.

In the tiny *Leo* between 1850 and 1857, McMullen sailed 8222 miles in cruises between the Thames and Land's End. His next boat was *Sirius*, a 32 ft gaff cutter with bowsprit. In this he sailed farther afield in home waters, including a cruise right round Britain in 1863 – the first such yacht cruise in a small vessel. His next boat, the *Orion*, was again larger and built by Inman of Lymington, Hampshire in 1865. She was LOA 42 ft, beam 10 ft 5 in and draught 7 ft. She was very much a deep narrow English cutter of the period. On this there were several hands. On her first cruise to Devon and Cornwall, she narrowly escaped being driven ashore during a gale in the night when anchored with many other yachts off Torquay, for the local regatta. A number of yachts managed to claw off the shore (22 August 1868) and only one yacht was wrecked. On the same night a number of small vessels were grounded and lost or damaged – a not uncommon occurrence in the days of coasting sail and yachts without the weatherliness and auxiliary power of today. In 1877, *Orion* was converted to yawl rig and McMullen made a single-handed cruise in the Channel, having dismissed his two paid hands who were lazy – and smoked so much that he thought it must enervate them. In 1891 at the age of 61 when sailing in mid-Channel in his 27 ft lugsail yawl *Perseus*, McMullen, alone on board, died of heart failure at the helm.

R. T. McMullen's *Leo* in the Channel on the night of 31 August 1857.

The circumnavigation of England and Wales for pleasure in the smallest yacht to date was by Edward E. Middleton in the gaff-rigged yawl *Kate* in 1869. *Kate* made her northern passage via the Bowling Canal and was LOA 23 ft, beam 7 ft and draught 3 ft 9 in and established that the cruising of very small craft was possible anywhere round British coasts.

For more than a hundred years since, numberless yachts of all shapes and sizes, large and small have cruised the waters of Europe, America and elsewhere and across the oceans. Many such cruises are in progress as you read these lines, as is obvious from the growing membership of the OCC mentioned above – and not all persons on such cruises, particularly non-English speaking, bother to make contact with the organization. Although there have been a great number of them, the number of long voyages with only one man on board – single-handed sailing – is much more limited. So are those by a crew of just two or in unusual craft. Since about 1960, the number of single-handed cruises anywhere each year has increased markedly; one reason for this is the development of effective vane steering which permits the yacht to maintain her course relative to the wind, while the lone mariner is resting or doing other jobs on board.

Single-handed voyages are better known and documented than two-handed cruises. This is because with two men, one or other is likely to change *en route* or even the crew members vary from port to port, with at one time or another perhaps two, three or four men or women on board. Single-handed voyages are by nature of a more set purpose: one might just add that the originators are often a shade more egotistical and more careful in providing records of their doings.

SHORT-HANDED VOYAGES

Two-man exploits

The first two-man west to east crossing of the Atlantic was made by a 43 ft ketch *Charter Oak* (skipper C. R. Webb) in 1857. As the prevailing winds in the North Atlantic are westerly, it is not surprising that this was the first direction for a short-handed passage.

The first two-man voyage from east to west was by the 20 ft converted lifeboat *City of Ragusa* in 1870. It was sailed by an American, John C. Buckley and an Austrian, Nicolas Primoraz. The yawl-rigged craft, with topsail and bowsprit and all inside ballast, took 84 days from Cork, Ireland to Boston, Mass.

The first two-man crew to cross the Atlantic from west to east in a yacht less than 30 ft was the iron ship-rigged (i.e. square sails on three masts) lifeboat *Red, White and Blue.* The crew were the Americans William Hudson and Frank E. Fitch. They left New York on 9 July 1866 and arrived at Deal, England 35 days later. The boat was LOA 26 ft, beam 6 ft 1 in. The feat was to publicize a new sort of 'metallic lifeboat' and it attracted attention from the Press of the day. McMullen commented on it as an unusual event. After the voyage the boat was put on exhibition in Paris and in London at the Crystal Palace.

The first man and woman crew to cross an ocean and to cross the Atlantic from west to east were Mr and Mrs Thomas Crapo. They sailed from Cape Cod to Penzance in 1877 in the 20 ft whaler *New Bedford*.

The first two-man crew to round Cape Horn from west to east (the direction of the invariably very strong prevailing winds) was on board *Pandora*, a sloop of 37 ft and a replica of Joshua Slocum's *Spray* (p. 155). This was on 16 January 1911 and was by an Englishman, G. D. Blythe and a Greek hand, Peter Arapakis.

The first two-man crew to circumnavigate the world were William Albert Robinson (USA) and Etera, a Tahitian. For the first part of the voyage from New York via Bermuda and Panama to Tahiti the second man was Willoughby Wright, but Etera stayed with Robinson, an American for the rest of the three-year cruise through the South Seas to Singapore, via Ceylon (Sri Lanka), the Red Sea, Mediterranean, Gibraltar, the Canaries and back to New York. The voyage lasted from 10 June 1928 to 24 November 1931, 3 years $5\frac{1}{2}$ months, distance 32 000 miles. The yacht was the Bermuda ketch *Svaap* (32 ft 6 in, beam 9 ft 6 in, draught 5 ft 6 in), designed by John G. Alden, a well-known American designer. These proportions are most moderate and not extreme in any way. *Svaap* had a 10 hp Kermath auxiliary motor and about 200 books on board.

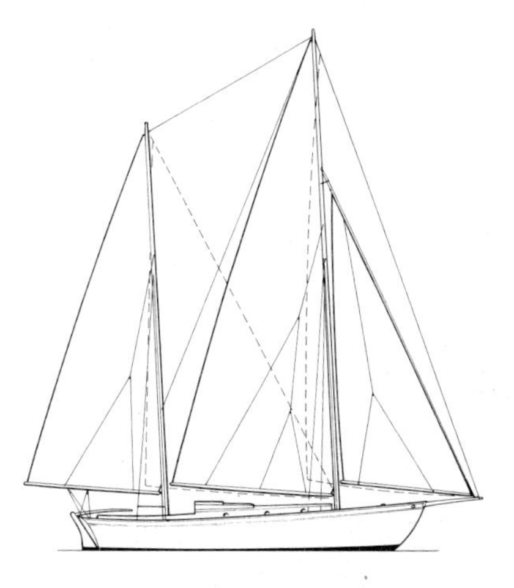

Svaap, the 32 ft 6 in ketch which was the first yacht to circumnavigate the world with a two-man crew.

The first woman to sail round the world as part of a two-man crew was a Mrs Strout (USA) who with her husband, Professor Strout, circumnavigated in another replica of *Spray*, *Igdrasil* in 1934–7. These replicas of the *Spray* were not uncommon in America. Apart from the thought that the shape of a yacht which had made such voyages must be seaworthy, *Spray* was uniquely known to have sailed herself off the wind with the helm lashed and this was a considerable attraction.

Left
Bruce Webb and Hugh Welbourn on the completion of their voyage round the world in 1974.

The schooner *Gazelle* entering Portsmouth on 9 May 1974 after sailing two-handed round the world.

The fastest two-man voyage round the world was in the 47 ft schooner *Gazelle* by Bruce Webb and Hugh Welbourn (Brit.) in 1973–4. She was designed by John Illingworth and built by CMN of Saint-Malo, France. The yacht left Portsmouth on 8 September 1973 (following the Round the World race but not permitted to enter because a full crew was not carried) and the plan was to use self-steering vane gears. *Gazelle* sailed to Cape Town in 54 days, distance 8292 miles and logged 6·4 knots, the highest known speed for a two-handed ocean passage. Among extreme weather met by the pair was a squall of 60 knots, which laid them flat when 1000 miles short of Cape Horn. After clawing off their sails and others being ripped to pieces, *Gazelle* surfed at 8 to 10 knots under bare poles in an estimated 75 knot wind (the anemometer was jammed at 60). The seas were completely white with driving foam and spray. Later the wind eased to 35 to 40 knots, only to go to 60 knots again 'accompanied by torrential rain that set the sea smoking and visibility only 100 yards or so'. She arrived back at Portsmouth on 9 May 1974 having taken 243 days on the whole round the world voyage.

The first two-woman ocean crossing was by Joan Baty and Stephanie Merry (Brit.) in a 32 ft Bermuda sloop in 1974. They left the Island Sailing Club, Cowes, on 17 April, crossed the Atlantic and anchored off the Seawanhaka Corinthian Yacht Club in Long Island Sound 5 weeks and 4 days later. Apart from a broken shroud, which was repaired and a toilet which went out of action, the trip was without serious incident.

Unusual cruising exploits

The greatest number of crossings of the Atlantic, mainly between England and the West Indies, including the northern route to the USA, by anyone in a small yacht is thought to have been made by Humphrey Barton (Brit.). By the end of 1974 he had completed 19 crossings and was planning the twentieth.

The most deliberate series of voyages of hardship and adventure under sail since 1945 have been undertaken by Major H. W. 'Bill' Tilman. Though born in 1898, he made a number of voyages from 1955 onwards to remote places including the Arctic and Antarctic, with a crew of young men. On arrival at a suitable place, they engaged in mountaineering often in snow and ice. Three boats have been used for these voyages, all Bristol Channel pilot cutters built around the turn of the century and between 45 ft and 50 ft. Tilman spurns modern yachting techniques and comforts. The first yacht *Mischief* was lost in 1968 off Jan Mayen, Iceland when under tow. The second old pilot cutter foundered south of Angmagssalik, Greenland. In 1955–6, Tilman took *Mischief* to the coasts of South America and the Strait of Magellan in the Cape Horn region. In 1963 the voyage was to the bleak shores of Baffin Bay, in 1967 it was to Punta Arenas and the South Shetlands. The yacht is the only one known to have inspired official geographical names – Mont de Mischief by the French authorities on Ile de Possession, Iles Crozet and Cap Mischief on Ile de Kerguelen; also Mount Mischief by the Canadian authorities on Baffin Island.

In 1972–3, David Lewis (Brit.) sailed from Australia in the 32 ft *Icebird* and sailed round part of Antarctica and back (see page 162).

The most remarkable survival of a cruising yacht is that of the 46 ft ketch *Tzu Hang*. On 14 February 1957 on a voyage from Melbourne for Cape Horn and thence to England, the yacht was hove down and pitch poled by a huge sea, which swept away rudder, masts, rig and all the deck structure and left a gaping chasm in the deck. The crew of three, Brigadier Miles Smeeton, DSO, MC, his wife Beryl and John Guzzwell (owner of *Trekka*, p. 158) set to work to survive in the sinking yacht. Covering up the holes with canvas in the terrible gale conditions of the Roaring Forties and making a jury rig, they eventually made a Chilean port after six weeks. Nine months later after an extensive refit they were again approaching the Horn, some 300 miles to go, and were lying a-hull in a severe gale to obviate the possibility of again pitch poling. But a great sea hove them down and the rig was again swept away. This time without John Guzzwell

A deck scene on *Tzu Hang* in the Roaring Forties in a lonely condition. Note huge swell and reefed mainsail.

they patched up the vessel and struggled northward to Valparaiso. A lesser couple would likely have given in against such rigours at sea, but their experiences of near total loss can explain some of the cases of yachts and crew which have disappeared without trace. The Smeetons in *Tzu Hang* actually rounded Cape Horn on a later occasion (in December 1968) from east to west.

The longest cruise in an open boat was by David Pyle (Brit.) and one crew. This was an 18 ft unballasted dinghy, *Hermes*, of the Drascombe Lugger class which sailed from Chichester, England to Darwin, Australia, leaving on 27 April 1969 and taking 341 days. The route was through the canals of France, through the Mediterranean to Mersin in southern Turkey, thence by truck and train to the northern reaches of the Tigris in Iraq. Pyle then sailed the length of the Tigris, 1000 miles, possibly the first such feat by a European, to Shatt-al-Arab and the Persian Gulf. The voyage was then along the coasts of southern Asia to northern Australia. The boat sailed 10 000 miles with extra mileage by overland transport and 230 days were spent on sea, river and canal.

The 'vertical' sailing record is held by Tristan Jones (Brit.). He took the 38 ft yawl *Barbara* to the Dead Sea, 1250 ft below sea-level, in 1970 and then sailed round the Cape of Good Hope to the West Indies. There he changed boats to *Sea Dart*, a 20 ft sloop with keel and bilge keels. He arrived at Lake Titicaca, 12 500 ft above sea-level, on 4 January 1974. This was via the Panama Canal and against the prevailing wind and Humboldt Current to Lima, whence it was trucked over the Andes to the lake, actually attaining 14 500 ft, the highest place to which a seagoing craft has ever been taken.

On the voyage in *Barbara* from the Cape to the West Indies, Jones sailed farther up the River Amazon than any previous sailing-craft, 1200 miles. He also rested at Devil's Island, the former French penal colony.

A number of unconventional craft have made extensive ocean passages. 'Unconventional' covers anything which is not designed to sail reasonably fast for its size to keep the trip within a reasonable time; or without the comforts of a cabin and proper shelter. Rafts of wood, rubber and other substances have been sailed and boats have been rowed (see p. 179). Very small craft come into the unconventional category, though some would say any ocean voyage in a yacht is unusual. Before 1945 there were few such odd trips recorded. Strangely some of the earliest single-handed voyages were in boats under 15 ft. Josiah Lawlor (USA) in *Sea Serpent* and William Andrews (USA) in *Mermaid* were both 15 ft and raced from Boston in 1891 to Coverack, England. The former made it in 45 days; the latter was rescued. A year later Andrews sailed in a slightly smaller craft, *Sapolio*, a chine canvas-covered craft, from Atlantic City to Palos de Maguer, Spain in 84 days. Bernard Gilboy (USA) took a 19 ft sloop from San Francisco to Australia in 1882, being rescued just before arrival. In 1928 Franz Romer (Ger.) sailed a 21 ft folding kayak from Lisbon to the Antilles (beam 3 ft; draught 1 ft), but later disappeared at sea on the way to Florida.

Single-handed exploits

Up to the end of 1974 there are thought to have been some 160 voyages across oceans by single-handed sailors. Remember there are borderline cases which are not across oceans and undoubtedly unreported crossings, by persons who do these things to get away from civilization and resist communicating as part of their whole philosophy. The number though, as already pointed out, is limited as compared with the thousands of yacht cruises and passages in small craft made all the time. Only 34 of these known voyages took place before 1945. The great enthusiasm for such exploits took place after the Second World War. This was due like other expansion in sailing to the pressures of modern life combined with increasing affluence; more tendency to 'opt out' (though in a most tenacious way!);

A contemporary picture of Joshua Slocum rowing *Spray* out of Yarrow River, Melbourne.

for single-handing in particular, the development of excellent vane gears for steering a boat on most points of sailing without the attention of a helmsman (and without power, i.e. wind-operated equipment); and the arrival in 1960 of the single-handed transatlantic race. This increased the numbers in actual participation and encouraged a wider public to take an interest.

There are four immortal firsts in single-handed ocean sailing.

The first man ever to sail across an ocean single-handed was an American, Alfred Johnson, in 1876, the 100th anniversary of the American Revolution. The little craft was, therefore, called *Centennial*. She was a 20 ft dory with gaff cutter rig, flush decked and with a canvas cover for the cockpit. The boat was unballasted and once turned over for a period of 20 minutes until she could be righted. The voyage was 64 days from Gloucester, Mass. to Abercastle, Wales, where Johnson arrived on 18 August – average speed 2 knots.

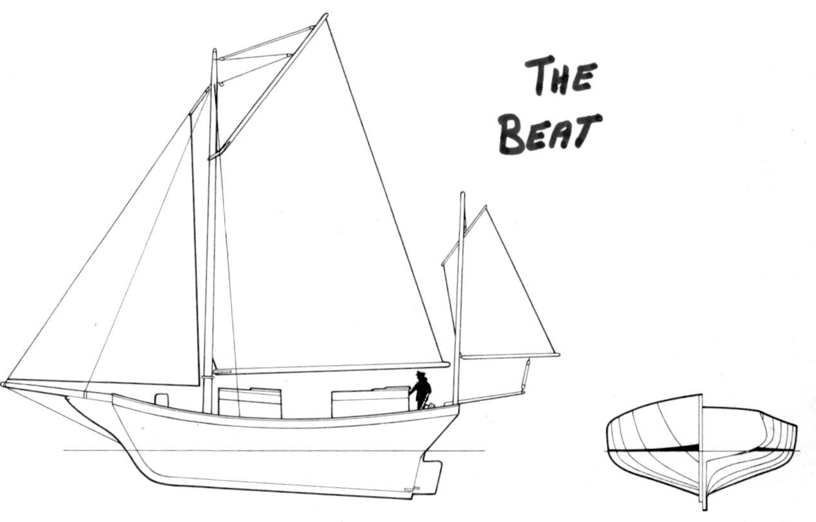

The profile and lines of *Spray*. The mizzen was put on during the voyage. Because of her remarkable self-steering qualities a number of replicas were built from time to time for ocean voyaging.

The first man to circumnavigate the world alone was Joshua Slocum (USA), a professional seaman. His famous yacht was a sloop *Spray*, converted during the voyage to a yawl with a small mizzen. She was LOA 36 ft 9 in, beam 14 ft 2 in, and draught 4 ft. Slocum built her by moulding – that is exactly copying by using as a mould – a rotten hull of the same name which he thought had been an oyster-boat on the Delaware at the end of the eighteenth century. 'Her lines were supposed to be those of a North Sea fisherman.' However Slocum added a foot of freeboard to the original model. The unprecedented voyage began when Slocum left Boston at midday on 24 April 1895 and ended when he anchored there again at 1 am on 27 June 1898. He had sailed alone round the world via the North Atlantic, Gibraltar, South America and the Magellan Strait, the South Pacific, eastern Australia, the Indian Ocean, Cape of Good Hope, West Indies and back to the east coast of the USA. He was 54 when he completed the voyage. After his return he continued to sail *Spray* on long cruises such as to the West Indies and back. In 1909 he set out single-handed from Bristol, R.I. for the Orinoco, but was never seen again. He is believed to have been run down by a steamer, probably when asleep below. Arthur Ransome once wrote 'Captain Slocum's place in history is as secure as Adam's.'

Joshua Slocum

The first man to sail round the world without calling in at any port was Robin Knox-Johnston (Brit.). He left Falmouth, Cornwall on 14 June 1968 and arrived back there without having touched land on 22 April 1969, 313 days out. Average speed 3·6 knots. He had sailed down through the Atlantic, round the Cape of Good Hope, through the Roaring Forties south of Australia and New Zealand, round Cape Horn and back up the Atlantic to England. His yacht *Suhaili* was a Bermuda ketch, built in Bombay, India, in 1964 and sailed back from there by Knox-Johnston. Its dimensions were LOA 32 ft 5 in, LWL 28 ft, beam 11 ft 1 in, and draught 5 ft 6 in. A specially contrived self-steering gear was fitted, just forward of the cockpit. Knox-Johnston was just over 30 when he completed his voyage.

The first man to sail alone round the world against the prevailing winds, that is from east to west in the Roaring Forties, was Chay Blyth, OBE, BEM (Brit.). In the specially designed and built 59 ft (LWL 43 ft 6 in) steel ketch *British Steel*, he left Hamble, England on 18 October 1970, headed south for Cape Horn which was rounded on Christmas Eve. The big steel ketch then smashed to windward through the Roaring Forties. This would have been impossible in former times, but was now feasible because of the development of modern ocean-racing techniques in sailing against the wind. (The square-riggers of the previous century sometimes gave up trying to round the Horn to the westward and ran back round the world the other way to Australia.) Blyth passed Cape Town on 25 May 1971 and returned to Hamble on 6 August 1971 (average speed 3·85 knots) where he was welcomed by members of the Royal Family and the Prime Minister. Blyth, who was 30 when he started out, also broke the record for the longest single-handed yacht ever to sail round the world. The extreme hardship of the endeavour means it will seldom be repeated. The voyage is the fastest single-handed east to west one.

These four 'firsts' are outstanding and on their own. Each feat was considered quite impossible by other sailors before it was done. All the men were of great experience – three were professional sailors previously and Blyth had rowed across the Atlantic (another 'impossible' feat). A fifth comparable voyage might be a single-handed circumnavigation by a woman, but this has not been attempted. (A woman has sailed across the Atlantic alone, see p. 165)

The second man to sail alone round the world was Harry Pidgeon (USA) in the hard chine 35 ft yawl *Islander*. The boat was built from plans made available to the public by a boating magazine. He sailed round via the Panama Canal on a 4-year voyage between 1921 and 1925. He made a second circumnavigation between 1932 and 1937, also in *Islander* and was the first man to have circumnavigated alone twice.

The first European to circumnavigate single-handed was the Frenchman Alain Gerbault. In the 36 ft ex-racing cutter *Firecrest*, he had an exceptionally stormy crossing of the Atlantic. He was the first man to sail alone across from east to west (Slocum sailed west to east and then south) then sailed by way of the Panama Canal to the South Seas where he spent much time. The voyage took 6 years 3 months and he returned to France in July 1929. Later Gerbault returned to the Pacific in a new boat and died there in 1941.

Suhaili leaves Falmouth.

The first single-hander to round Cape Horn was Al Hansen (Norw.) in the 36 ft cutter *Mary Jane* in 1933 from east to west, but he was lost with his yacht when she was wrecked soon afterwards.

Robin Knox-Johnston on *Suhaili*.

The first Englishman to cross an ocean single-handed was Commander R. D. Graham, RN (rtd). He left Bantry, Ireland on 26 May 1933 arriving at St Johns, Newfoundland 24 days later. This fast passage across the Atlantic was made in a conventional 30 ft five-year-old gaff cutter, *Emanuel*. The route taken was farther north than any previous lone voyages; although colder and with danger from icebergs, Graham expected a shorter distance and more favourable winds north of the customary west-going depression tracks. In 1939, Graham set out to sail round the world accompanied by his daughter in a larger yacht, *Caplin*. He sailed from England to New Zealand, arriving there in November, but as the war had begun the voyage was terminated.

The only single-hander to sail round the world during the Second World War was the Argentinian Vitto Dumas in the 32 ft ketch *Lehg II*. He sailed from Buenos Aires to Cape Town, round the southern ocean south of Australia to Wellington, NZ, then Valparaiso and on south of Cape Horn, and back home in 272 days between July 1942 and August 1943. His route was, therefore, clear of combatant areas. He was the first single-hander to round Cape Horn and survive and the first ever from west to east.

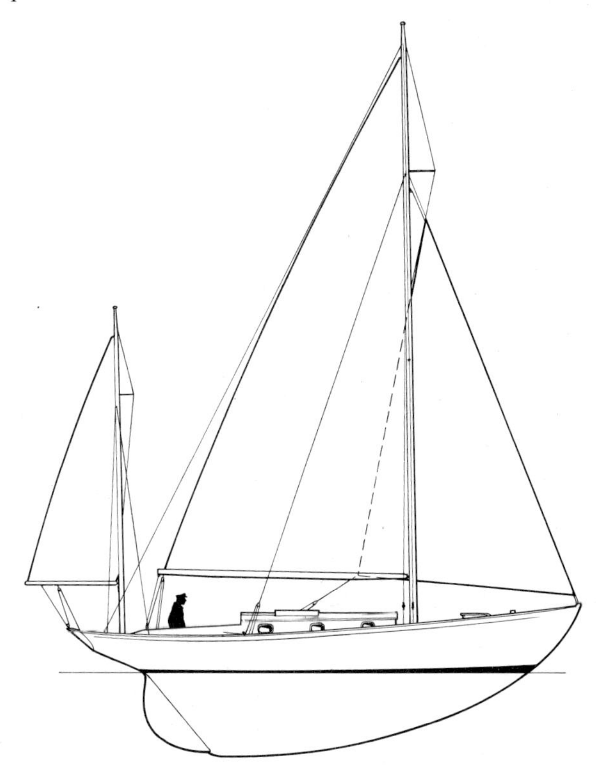
Stortebecker III, the 33 ft German yawl that was the first to cross an ocean using the Bermuda rig.

Before 1945 the fastest voyage by a single-hander round the world was that by Louis Bernicot (Fr.) in his 41 ft Bermuda cutter, *Anahita* via the Magellan Strait in 1 year 10 months. He arrived back in Carentec, France on 30 May 1938 having left there on 22 August 1936. The Bermuda rig had only come into general use for racing boats in the 1920s and the first use by an ocean voyager was by Fred Rebell (Latvia) on the 19 ft yawl *Elaine* in the Pacific. The first crossing of the Atlantic with such a rig, which was then widely regarded as 'unseaworthy', was by *Stortebecker III*, an efficient 33 ft yawl sailed single-handed by Ludwig Schlimbach (Ger.).

John Guzzwell of *Trekka* and *Tzu Hang*.

Exploits from 1945 to 1960

The period from the end of the Second World War to 1960 is a convenient period to review, because 1960 was the start of the first single-handed transatlantic race. This suddenly made such voyages seem less ordinary and increased their popularity as well as actually increasing the numbers recorded because of the entries in the competition itself. The publicity, too, was greatly increased and encouraged lone exploits and sponsorship of them. Up to 1960 the possibility of races or speed records was little considered: to get to the other end was an achievement.

In the 15 year period lone circumnavigators were five in number *Stornaway*, 33 ft cutter, sailed by Alfred Peterson (USA), via Panama Canal, June 1948 to August 1952. *Atom*, 30 ft yawl, sailed by Jean Gau (Fr.), via Panama Canal, September 1953 to October 1957. *Le Quatre Vents*, 31 ft cutter, sailed by Marcel Bardiaux (Fr.), via Cape Horn, but a drawn-out cruise of over 8 years, May 1950 to July 1958. *Kurun*, 33 ft cutter with gaff main, topsail and bowsprit and Colin Archer-type hull, sailed by Jacques-Yves le Toumelin (Fr.), via Panama Canal, from Le Croisic, France and return, 19 September 1950 to 7 July 1952 (he did, though, have companions part of the way). *Trekka*, 20 ft Bermuda yawl with high-aspect ratio and fin keel and separate rudder purpose designed by Laurent Giles (Lymington, England), sailed by John Guzzwell (Brit.) between September 1955 and July 1959, Vancouver, Canada via Panama Canal and back. During the journey, he left the boat to join *Tzu Hang* and took part in her first capsize and survival.

During the period 1945 to 1960 approximately 60 single-handed yachtsmen made ocean cruises.

Unusual craft to have been sailed during this time were: *Heretique*, a 15 ft rubber raft, in which Dr Alain Bombard (Fr.) drifted and sailed from Casablanca to Barbados between 24 September 1952 and 22 December 1952. His purpose was to try survival techniques especially living off fish and natural organisms in the sea. He carried *no food and water*. This is the only deliberate survival experiment made across the Atlantic or any other ocean. The only propulsion in the U-shaped air-filled craft was a short mast. There were leeboards and a sea anchor which could be lowered to stop excessive drift. Bombard, aged 27 at the time, succeeded in proving that a man could survive living on the sea. His experiment has been of great value to several others who found themselves in the same situation *involuntarily*. Just the knowledge that this remarkable Frenchman concluded his Atlantic crossing of 65 days living off fish and plankton added to their wills to survive. The 65 day stretch was from Las Palmas to Barbados. He had stopped at Las Palmas after an initial 12 day drift from Casablanca. But the second leg was 2900 miles without help, except for boarding a British ship on one occasion, after being persuaded on board for a light meal.

In 1954, William Willis (USA) then in his seventies and against all advice journeyed 6700 miles across the Pacific from Callao to Pago Pago on the metal and wood raft, *Seven Little Sisters*. It was 33 ft with house and lug mainsail and small mizzen. In 1963 he repeated this feat on another raft, *Age Unlimited*, this time from Callao to Australia.

Dr Hannes Lindemann (Ger.) crossed the Atlantic in 65 days in 1955 from Las Palmas to the Antilles, St Croix, a distance of 3000 miles. He was alone in the 17 ft collapsible kayak *Liberia II* with a 50 ft² single sail on a short mast.

Unusual craft after 1960

Another raft rigged as a cutter, 25 ft overall, was sailed by René Lescombe (Fr.) from Guadeloupe to Flores between March and June 1963, but subsequently disappeared at sea.

Tinkerbelle – a sloop measuring 13 ft long, beam 5 ft and draught 1 ft – sailed by Robert Manry (USA), a journalist, crossed from Falmouth, USA to Falmouth, Cornwall, England in 78 days in the summer of 1965.

The smallest yacht ever to make an ocean crossing – and therefore by definition because not more than one man could be carried, single-handed – was the *April Fool*, a sloop of 5 ft 11½ in. Sailed by Hugo S. Vihlen (USA), the boat left Casablanca on 29 March 1968 and arrived at Fort Lauderdale, USA on 21 June 1968, an 85-day passage over a 3000 mile distance, average speed 1·4 knots. Also in 1968 William Willis (USA) at the age of 77 started across the Atlantic from New York towards England on the 12 ft raft, *Little One*, but foundered at sea and was lost.

The smallest yacht ever to make an ocean crossing, the *April Fool*, 5 ft 11½ in, sailed by Hugo S. Vihlen (US).

Half Safe, named by its owner Ben Carlin who had no illusions about its suitability, was an 18 ft landing-craft for army use known as a 'DUKW'. The design was intended to be lowered off ships for military assault on beaches. Carlin and his wife sailed this unusual craft round the world. The fuel reserve was lashed in cans on the structure. It could, of course, travel by road as well as sea to shorten ocean distance. *Half Safe* crossed the Atlantic from Halifax, Nova Scotia to Spanish West Africa in 1950–1 and then 'drove' to England. Subsequently they cruised round the world.

The smallest craft to have been sailed across the Atlantic and beyond by an Englishman was *Sea Egg* (or *Sjø Ag*), a 12 ft bilge-keel sloop shaped rather like an egg above water. She was built by the owner-voyager John C. Riding at La Rochelle, France and then sailed to Plymouth where entry for the 1964 transatlantic race was refused as the *Egg* was below minimum size. Riding was 6 ft 4 in and weighed 16 stone. He left Plymouth on 19 June 1964 and then called at Corunna and Vigo and left the Azores on 23 April

1965, arriving in Bermuda on 28 June. From there he sailed to Panama, Mexico and then to San Diego. On a further voyage across the Pacific he reached New Zealand, but on passage to Australia became well overdue and is believed to have foundered in a hurricane.

Exploits after 1960

The next period of 15 years has seen single-handed ocean voyaging becoming more common, the number of known circumnavigations of the world quadrupling over the equivalent previous period. Those who have completed this epic voyage are Bill Nance (Austral.), Sir Francis Chichester, Sir Alec Rose, Robin Knox-Johnston, Nigel Tetley, Rusty Webb, Chay Blyth, Bill King, Tom Blackwell, Graham Dillon (Brit.), Jean Gau (for the second time), Leonid Teliga, Bernard Moitessier, Roger Plisson, Alain Colas (Fr.), Walter Koenig, Alfred Kallies, Frank Casper, Wilfried Erdmann, Rolla Gebhard, Joergen Meyer (Ger.), Alan Eddy, John Sowden, Robin Lee-Graham, George Darling (USA), Krzysztof Baranowski (Pol.), Kenichi Horie (Jap.), Michel Mermod (Switz.). The small number of nationalities concerned is remarkable, since yachts of numerous other countries cruise the seas. Among these circumnavigations are the following with special features. The first man to sail alone round the world with only one port of call was Sir Francis Chichester (Brit.). Before his voyage it had been customary to cruise and put into numerous ports. He remains the only person to have undertaken this specific feat. His intention was to beat known record times of the clipper ships, but probably this was asking too much for a single-handed venture that had never even been tried before. This was not achieved, but, the voyage remains the fastest single-handed west to east rounding of the world by a single-hulled yacht. The yacht was the 54 ft ketch *Gipsy Moth IV*. She was designed by Illingworth and Primrose and built by Camper and Nicholsons, Gosport, England. Chichester left Plymouth, England on 27 August 1966 and called at Sydney, Australia only. He arrived back in Plymouth on 28 May 1967. At the time *Gipsy Moth* had made a passage longer than any previous ocean voyage by a sailing-yacht: 15 500 miles from Plymouth to Sydney. This took 107 days, 6·03 knots. From Sydney round Cape Horn to Plymouth was 14 500 miles and took Chichester 119 days (5·07 knots). The total number of days at sea was 226, average speed 5·71 knots. Chichester was the first man to be commercially sponsored on a long voyage (as opposed to actually taking on a voyage to prove a design or special boat) and carried the 'Wool mark' on his bows and own hat! This was of significance in the expansion of ocean voyaging, setting a precedent for the financing of future feats.

Sir Alec Rose (Brit.) followed Chichester with the intention of stopping only at Melbourne, but was forced to put in to Bluff, New Zealand for repairs but his yacht, the 36 ft *Lively Lady*, was smaller and older than *Gipsy Moth* and he had little financial support. He left Portsmouth on 16 July 1967 and returned there on 4 July 1968. His was, therefore, the only yacht under 40 ft to have sailed round the world alone with two stops or less and the second yacht of any size to have done this. He was also the second grocer in history to be elected a member of the Royal Yacht Squadron (the first was Sir Thomas Lipton).

The first person to have sailed round the world single-handed without any port or point of call was Robin Knox-Johnston, CBE (Brit.) in *Suhaili* (see p. 156).

Gipsy Moth IV sailed alone by Francis Chichester: she was the first yacht to encircle the world calling at only one port.

Alec Rose sets out on *Lively Lady*.

Bernard Moitessier

The first person and only one to have sailed round the world single-handed east to west (against the prevailing winds of the southern ocean) was Chay Blyth, CBE, BEM (Brit.) in *British Steel* (see p. 156).

The longest single-handed passage without putting in at any port was by Bernard Moitessier (Fr.) who left Plymouth in his 39 ft 6 in all-steel Bermuda ketch *Joshua* (named after Joshua Slocum) (beam 12 ft, draught 5 ft 3 in) on 22 August 1968. The voyage was intended to be round the world, but returning to England (as part of the world single-handed race, see p. 171). He sailed down the Atlantic, rounded the Cape of Good Hope and sailed eastward in the Roaring Forties south of Tasmania and New Zealand and rounded Cape Horn (he had done this previously) in February 1969. He then headed north for Europe but on 18 March passed a message by catapult to the bridge of a ship: 'The Horn was rounded February 5 . . . I am continuing non-stop towards the Pacific Islands, because I am happy at sea, and perhaps also to save my soul.' He passed south of Australia, an unprecedented second time, in April 1969 and finally put into Tahiti in the storm-beaten *Joshua* on 21 June 1969. On arrival he said, 'Talking of records is stupid, an insult to the sea. The thought of a competition is grotesque. You have to understand that when one man is months and months alone one evolves; some say people go nuts. I went crazy in my own fashion. For four months all I saw were the stars. I didn't hear an unnatural sound. A purity grows out of that kind of solitude. I said to myself: What the hell am I going to do in Europe? I told myself I would be crazy to go on to France!' When he arrived, he had not been heard of or seen for three months. He had been at sea on his own a continuous 301 days – the longest by any solitary sailor on passage and an immense and record 37 455 miles. No other yacht or small sailing-vessel or single-hander has ever spent eight consecutive months in the Roaring Forties – covering 29 000 miles! *Joshua* was twice knocked down (mast head touching close to water) in the Indian Ocean and twice in the Pacific, but the strength of her steel hull saved her from a fate like that of *Tzu Hang*. The average speed of *Joshua* was 5·18 knots.

The longest single-handed passage of all. The steel ketch *Joshua* at sea under shortened sail.

The first man to sail single-handed to Antarctica and to attempt to encircle it was Dr David Lewis (Brit.-NZ) in the 32 ft steel sloop *Icebird*. He left Sydney on 19 October 1972 and Oban on the south coast of New Zealand on 2 November. Most of the voyage was between the 60th parallel and the farthest north pack-ice. Although midsummer (January) the weather was bitterly cold and there were almost continual gales. On 29 November Lewis (who in other boats had sailed round the world with his family, three times crossed the Atlantic single-handed, and also sailed to the Arctic in a multi-hull) was in hurricane force winds of 70 knots and the worse conditions that he had ever experienced. 'The whole sea was white . . . these are not

Susan Hiscock and her husband Eric have made a series of very extensive cruises beginning in 1952 in the 30 ft cutter *Wanderer III* with a circumnavigation of the world. The two of them later made a second round the world voyage, then over the years trips across the Atlantic to the North American Pacific coast. In 1968 they set off in a larger yacht, *Wanderer IV*, again for the Pacific, returning to England in June 1975 for a period between voyages.

In the Whitbread Round the World race Wendy Hinds completed the entire course as one of the crew of 13 on board the 71 ft *Second Life*; other women sailed on various legs across the oceans on this venture. They have sailed in demanding races like the Round Britain (two in crew maximum) and, of course, are continually in yachts on cruises, long and short, everywhere.

Few women have, however, sailed long passages single-handed. The reasons are mainly concerned with sheer strength required for looking after a boat at sea without help, not to mention problems for a woman by herself in a foreign harbour and differences from men in attitude to 'roughing it' – without much washing.

The first woman to cross an ocean was Anne Davison (Brit.) in the 23 ft sloop *Felicity Ann*. She left Plymouth on 18 May 1952 and reached Miami, Florida on 13 August 1953, later sailing on to New York. The yacht was small enough to be easily handled by the lone female sailor with a correspondingly small sail area and so was slow. Her LWL was 19 ft, beam 7 ft 6 in, draught 4 ft 6 in, and sail area 237 ft^2. The yacht was a wooden one constructed by Cremyll Shipyard, Plymouth, work having begun in 1939 and completed after the war in 1949. The main part of the crossing was from the Canaries on 1 December to Dominica on 24 January, 3300 miles at 2·3 knots average. Mrs Davison had previously been wrecked in another craft with her husband off Portland Bill, on the way to the West Indies. She had managed to swim ashore after 14 hours mostly in the dark, but her husband, Frank, died of exposure. After this she resolved to sail on alone. After her arrival in the USA she settled there and married an American.

The only woman to sail single-handed across the Pacific is Sharon Sites (later Adams) (USA). Her first voyage was in the 25 ft sloop *Sea Harp*, from San Pedro, USA to Hawaii, distance 2300 miles making a 39 day passage and arriving on 20 July 1965. In 1969 she made a longer journey from Yokohama, Japan to San Diego, USA in *Sea Harp II*, 31 ft, a 5000 mile passage, of 74 days, arriving 24 July 1969.

Marie-Claude Faroux

The fastest crossing of the Atlantic by a woman was in 1972, when Marie-Claude Faroux (Fr.) sailed *Aloa VII* in the single-handed transatlantic race from Plymouth, England to Newport, R.I. in 32 days 22 hours 51 minutes (average speed 3·54 knots). The yacht was standard French Aloa class ocean racer (another one also sailed in the race) and was LOA 34 ft 7 in, LWL 24 ft 4 in, beam 11 ft, draught 5 ft 11 in. Mlle Faroux was already a national champion in the Moth dinghy class in France. When she reached Newport after her crossing a report said '*Aloa VII* looked trim and well maintained, while the skipper was in striped jeans and pink sweater, her long hair neatly plaited and her teeth shining white as she smiled shyly to the well wishers.' In the last part of her trip she encountered much fog and was constantly in fear of being run down. At 26 she is also the youngest girl to have crossed an ocean alone.

Other women crossing the Atlantic at the same time were Teresa Remiszewska (Pol.) in *Komodor*, 42 ft yawl, who took 57 days and Anne Michailof (Fr.) in *P.S.*, 30 ft sloop, who took 59 days.

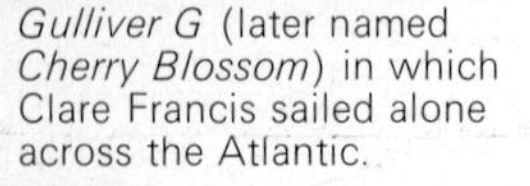
Gulliver G (later named *Cherry Blossom*) in which Clare Francis sailed alone across the Atlantic.

The second woman to cross the Atlantic, and to make the fastest crossing up to that time by a woman, was Nicolette Milnes Walker (Brit.), 27, in the glass-fibre Pioneer class sloop *Aziz*, 31 ft. She left Dale, South Wales on 12 June 1971 and reached Newport, R.I. in 44 days on 26 July, having covered a distance of 3400 miles (average speed 3·2 knots). Miss Milnes Walker took the 'intermediate route' across the Atlantic from east to west, sailing south from the British Isles and passing close to, but not stopping at, the Azores. Then she turned north-westward to reach the north-eastern seaboard of the USA. Anne Davison, however, took the southern route – that is with the trade winds – stopping at the Canaries and arriving in the West Indies. Only from there did she turn north to the eastern seaboard. She thus avoided the North Atlantic. Marie-Claude Faroux, who was racing and in a larger ocean racing boat designed essentially for beating to windward in all conditions took a northern route, though not the far northern short way, which might have met icebergs and very bad weather. As mentioned, she did meet fog in the later stages and this was not surprising on the chosen route.

In 1973, Clare Francis (Brit.) in the Nicholson 32 class glass-fibre sloop *Gulliver G*, crossed the Atlantic alone from Falmouth to Newport, R.I. in 37 days. This made her the sixth woman to cross the Atlantic single-handed. It is a fact that no woman has crossed the Atlantic alone from west to east (the direction of the prevailing wind in the North Atlantic) but this is presumably because all the female sailors have so far been Europeans and, therefore, sailed *to* America, rather than for any technical reason.

RACING LONG DISTANCES

The 'sport' of racing across an ocean single-handed is a new concept – it was not seriously considered before 1960. (There were matches in freak boats attempted in 1891 and 1892 from west to east, but no results.) Very few small yacht voyages had even attempted the east to west crossing of the Atlantic: the first single-hander on the route was Commander Graham

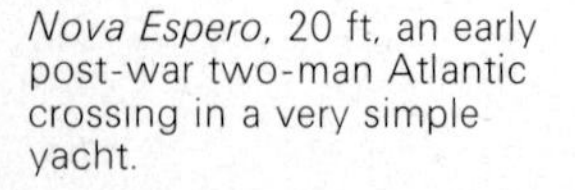
Nova Espero, 20 ft, an early post-war two-man Atlantic crossing in a very simple yacht.

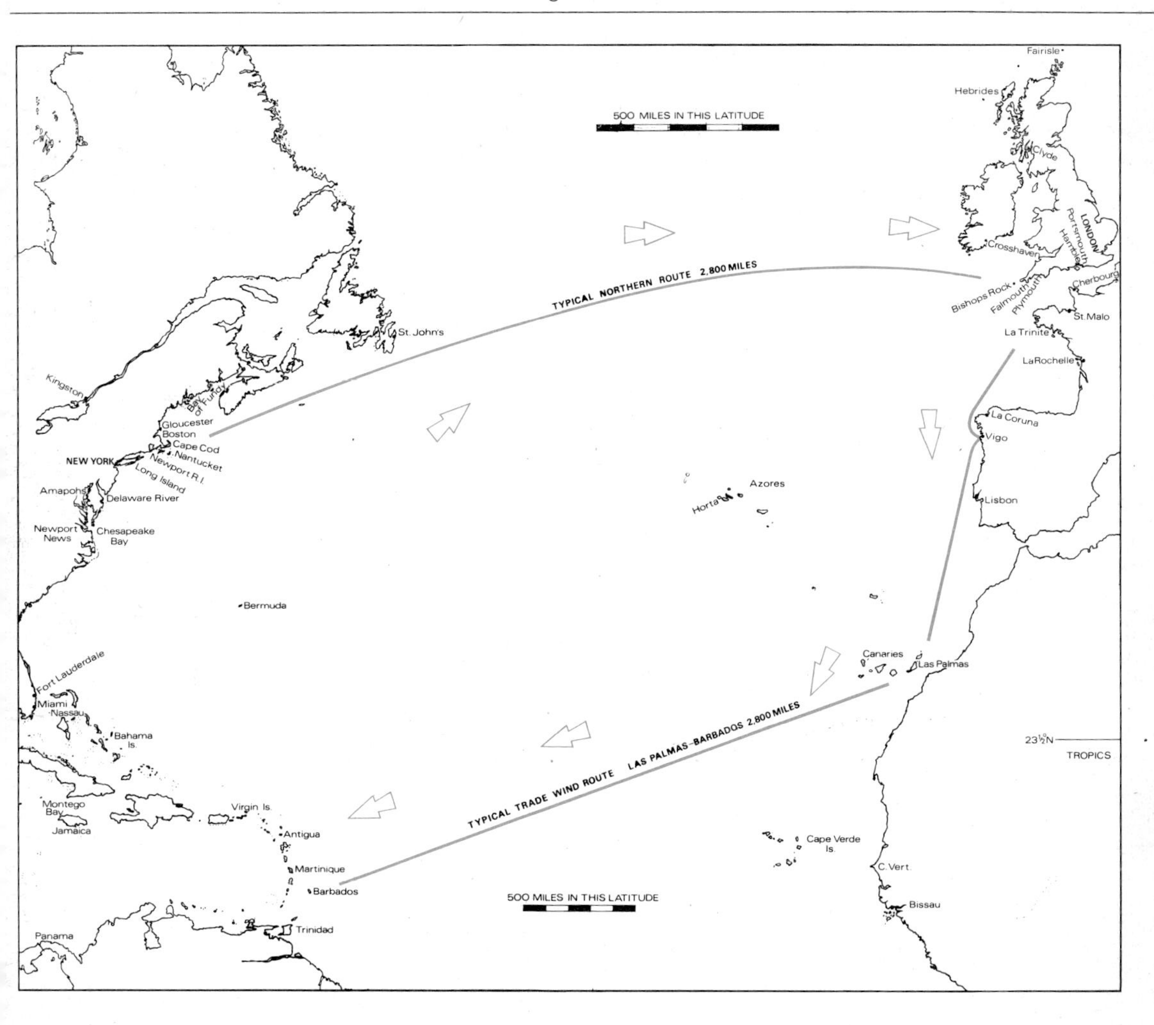

North Atlantic. Scene of hundreds of yacht voyages, showing common ports of call and landfalls for yachts. The northern route, commonly followed by transatlantic racers from east to west, but much windward work is expected The trade wind route is for more leisurely cruising, favourable winds are common. Total southern route is longer, but with the prospect of sunshine and ports of call.

(see p. 157) in 1933. Between 1945 and 1960 there were a few more including Humphrey Barton (Brit.) in *Vertue XXXV*, a 25 ft Bermudan sloop (1950) and *Nova Espero*, 20 ft gaff sloop sailed by Stanley Smith and Charles Violet (Brit.). Both these other passages were two-handed. One single-hander, Peter Hamilton (Brit.), crossed in a Vertue in 1956. The northern route from the British Isles or northern Europe to the eastern seaboard of the USA is the shortest in distance across the Atlantic, but the snag has always been – and still is – that it is against the prevailing westerly winds that drive on day after day urged by the depressions which cross the ocean. The prevailing ocean current goes with the wind. There are east winds at times, but they are infrequent.

What has changed the situation is the development in ocean-racing boats round the coasts of America, Europe, and Australia in weekend and longer races. By the mid-1950s yachts between 30 ft and 70 ft, fully crewed, were able to beat to windward for hour after hour in winds of 35 knots and more.

The old sailing-ships and early yachts found it more practical to do a 'Christopher Columbus' and sail to America with the trade wind (north-east) southward to the West Indies and only then turn north to the USA. Any yacht bound for a cruise and sunshine still does this. But if there is time to be saved or a race to be sailed then the northern route with its shorter distance beating to windward becomes a practical possibility. Of course, the clever sailor is not making a dead beat to windward all the time: he tacks to shifts, sometimes able to lay the course and sometimes not quite – and there are times of exception to the prevailing wind when the breeze frees and he can ease sheets or even hoist a spinnaker or other running sails.

Transatlantic

Up to the start of the first single handed transatlantic race the number of single-handed crossings of the Atlantic was about 20. The idea of a single-handed race was quite strange, for after all it was a feat of endurance to get there at all. It also implied dangers: the obvious one is that a lone sailor must sleep as the yacht sails on. He is not, therefore, keeping the look out which is the basis of good seamanship and the avoidance of collision. Since the race started there have in fact been collisions caused by this very defect, between yachts and ships and between yachts racing, though without serious consequences. Another danger is that of falling overboard with the steering vane set, so there is no one to sail back and pick up the lone skipper – there is the nightmare of seeing the yacht sail away while struggling in the water. Yachts pressing on with full crews, it was felt, was one thing but to put pressure on a man alone at sea to come on deck half asleep and trying to carry more sail or take risks to make the yacht go faster, was another.

Jester which has made seven passages between Europe and America: a Folkboat hull with Hasler's own Chinese lug rig.

The first man to put forward the idea of a single-handed sailing race was Colonel H. G. 'Blondie' Hasler, who had commanded the 'cockleshell heroes' in a raid by Royal Marines on enemy shipping up the Gironde in the Second World War. There was an aspect of the fully crewed and fairly expensive modern ocean yacht that Hasler wanted to outflank by means of the race. He felt that if the only important restriction was to limit the crew to one, then it would not be necessary to have complicated rating and handicapping, but instead a fairly moderately sized and easily handled craft would develop for the good of cruising yachtsmen as a whole. His idea inspired the race which was first held in 1960 and every four years subsequently (1964, 1968, 1972), but the competitors outsmarted his technical hopes with boats even larger than those permitted under customary ocean racing rules.

The only boat to have taken part in every single-handed transatlantic race from 1960 onwards is Hasler's original boat in which he sailed the first race, the 25 ft Folkboat class, *Jester*. This has a Chinese lug rig of special design using traditional Chinese principles but modern materials (Terylene, GRP, etc.) which obviates the usual requirement to clamber on deck, because it can be hoisted, trimmed, reduced in area and lowered all by the helmsman within the yacht with his head only exposed, or in a perspex dome. The skipper also steers from below and this means that he is safer and is protected from cold and exposure. In the second two races the skipper was Michael Richey, Secretary of the Royal Institute of Navigation, London. By the end of 1972, *Jester* had made 7 passages between Europe and America, more separate voyages on this route than any other boat of 25 ft or less.

The first yacht to win a single-handed race across the Atlantic was *Gipsy Moth III*, 37 ft, a wooden Bermuda cutter built in 1959. She was sailed by Francis Chichester, aged 59, and left Plymouth on 11 June 1960 arriving at the finishing-point, Ambrose Lightship, 3000 miles distance, after

40 days 11 hours 30 minutes. The other four men to take part in this first race was Blondie Hasler (Brit.) in *Jester* (2nd, 48 days), David Lewis (Brit.) in *Cardinal Vertue*, a 28 ft Vertue class sloop (3rd, 56 days – he broke his mast and put into Ireland), Val Howells (Brit.) in *Eira*, a 25 ft Folkboat (4th, 63 days after putting into Bermuda for repairs), and Jean Lacombe (Fr.) who took 69 days to cross, though an official 74 days because of having to put back to Plymouth for repairs.

The smallest yacht to take part in any of the single-handed transatlantic races was the glass-fibre Hunter class sloop *Willing Griffin*, 19 ft, LWL 17 ft 1 in, beam 6 ft 2 in, draught 3 ft 4 in, displacement 3300 lb. She was sailed by David Blagden (Brit.), aged 28, and made the crossing in 52 days 11 hours in the 1972 race.

In 1962, Francis Chichester, quite independently made a solo voyage across the Atlantic on the same course as the race in order to try and break the '30 day barrier'. This was the first of Chichester's attempts to make a 'solo race' to break a pre-set record. Other ones were his attempt to beat the average 100 days of the clippers to Sydney and the 4000 miles in 20 days sail in 1971. The 1962 trip was in *Gipsy Moth III* and she was sailed from Plymouth to Ambrose Lightship off New York in 33 days 15 hours.

The 1964 race attracted 16 starters, 14 of whom finished. It was the first one to be won by a Frenchman, the first one with multihull craft in it (three of them). A number of the competitors had arrangements with London papers to send back reports by radio and these were frequently published. However, Eric Tabarly, a Lieutenant in the French Navy, remained silent on the radio and concentrated on sailing his boat, the 45 ft ketch *Pen Duick II*, the largest yacht in the race. He finished first in 27 days 3 hours, thus easily beating the '30 day barrier'. (The race was to Newport, R.I. instead of New York but this was only about 120 miles shorter.)

The progressive times with winners and number of entries for the single-handed race are as follows:

Year	Winner	Yacht	LOA	Starters	Time	Average Speed
1960	Francis Chichester (Brit.)	*Gipsy Moth III*	39 ft	5	40 d 11 h	3·09 kn
1964	Eric Tabarly (Fr.)	*Pen Duick II*	45 ft	16	27 d 3 h	4·38 kn
1968	Geoffrey Williams (Brit.)	*Sir Thomas Lipton*	56 ft	35	25 d 20 h	4·60 kn
1972	Alain Colas (Fr.)	*Pen Duick IV*	70 ft	52	21 d 13 h	5·80 kn

The fastest crossing in the race is, therefore, the 21 days 13 hours 15 minutes by *Pen Duick IV*, a 70 ft trimaran sailed by Alain Colas. In 1960, 49 ft was considered the largest yacht that could be handled by one man but it has been shown that a very large yacht indeed can be sailed single-handed. The race is now invariably won by one of the giants. So handicap prizes and prizes for boats not exceeding a certain length have been presented. The next race is in 1976. The race is now sponsored by *The Observer*, the London Sunday newspaper. Hasler's original concept of the small, easily handled yacht has been to a great extent lost in the bid to be the first home in this event, but the race has become more and more popular as shown by the above figures. The intriguing question remains: How fast will a yacht sailed by one man be able to sail from Plymouth to Newport? It is certain that the existing records will be beaten, though in sailing this depends on good luck with the wind as well as hard sailing, stamina, and clever tactics. When there is a big yacht, unusually fair winds, and a determined helmsman the record will tumble again. An average of 8 knots which is very difficult but not impossible would be 15 days' sailing. (But remember the yachts sail much more than the nominal distance. Time goes on: average speeds remain low because little headway is made over appreciable periods against head winds and gales from ahead – when the the yacht can even lose ground.)

The fastest single-hulled yacht and the biggest ever to take part in the race and the biggest yacht to be sailed single-handed across any ocean was the 128 ft French three-masted staysail-rigged 'schooner' *Vendredi Treize* (Friday the Thirteenth) designed by Dick Carter (USA) and sailed by Jean Yves-Terlain. She was sponsored and owned by film producer Claude Lelouch who let it be known that he would follow his successful movie *A Man and a Woman* by one called *A Man and a Boat*. *Vendredi* was second boat home in 1972 and first single-hulled yacht in 21 days 5 hours. Other dimensions were LWL 116 ft 3 in, beam 18 ft 3 in, draught 11 ft 1 in; displacement 78 000 lb.

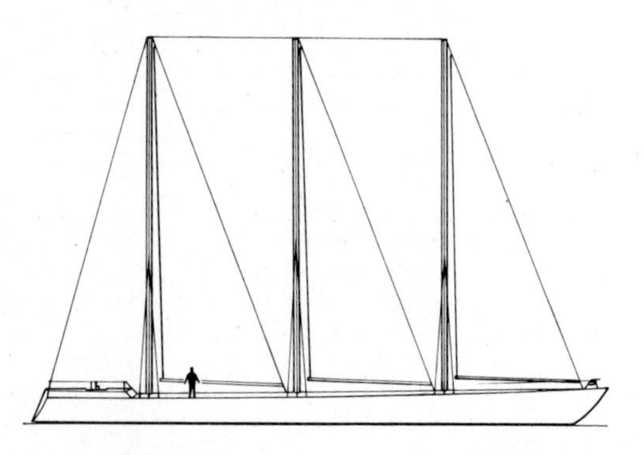

Vendredi Treize, largest single-hander.

The greatest number of multihulls to take part in any race was 8 in 1972. Of these 6 were trimarans and 2 were catamarans. Only three women have taken part in the single-handed transatlantic race: Marie-Claude Faroux, Teresa Remiszewska, and Anne Michailof.

The most criticism of the race and of its standard of safety came in the 1968 event. This was because of the number of yachts forced out by damage and the demand on rescue services. Eric Tabarly in *Pen Duick IV*, which was then new and unready, hit an anchored ship when he was below decks, returned to Plymouth for repairs but was unable to continue. Jean de Kat (Fr.) in the trimaran *Yashka* put into Alderney for repairs but then sailed on. Another French trimaran sailor in the race retired because he said his navigation relied on radio contact with Air France jets and the airline had gone on strike. The old ocean racer *Zeevalk* gave up with leaks and weaknesses. The only Swede ever to enter gave up because of the cold (?) and later sailed to the tropics. The trimaran *White Ghost* failed to get more than a few miles. *La Delirante* broke her mast. The Italian *San Giorgio* returned to Falmouth with a broken rudder. Another English yacht was dismasted through weak aerial insulators on the rigging. The French yacht *Silvia II* was dismasted at an early stage but later carried on and completed the course. The French *Ambrima* was dismasted and lost her rudder and taken in tow but sank. *Gunther III* retired with a broken mast step. In the later stages of the race, the skilled French skipper Alain Gliksman in *Raph* lost his rudder, but made Newfoundland with difficulty and without assistance. In mid-Atlantic the frail *Yashka* broke up and Jean de Kat took to his life-raft: luckily his distress call was heard by a civilian aircraft, but the RAF managed to locate this needle in a haystack, only with the aid of Service-type radar. De Kat was then fortunate to be picked up by a Norwegian freighter. The time and effort for this rescue culminated in considerable argument about the use of Services and commercial vessels and aircraft. Of the 35 starters only 19 made Newport, but after this regulations and inspections were tightened up to lessen the possibility of wholly unsuitable craft setting out. There has been no loss of life on any single-handed transatlantic race.

The wide increase in speed in the later races reflected the better sailing techniques and increased competition. It is not only the winners as shown above, which have made faster times. In 1972, the first 28 boats to finish had faster times than the winner of the 1960 race (40 days 11 hours).

Pacific

The only single-handed race across the Pacific was from San Francisco to Tokyo in 1969. There were 4 starters; 3 American and Eric Tabarly in *Pen Duick V* (35 ft, beam 11 ft, draught 3 ft). The latter was a specially designed sloop fitted with water ballast tanks, that is tanks that can be filled on either side, so that when the yacht is on a reach on with the wind from one side for a long distance she gets added power – the practice is banned in ordinary yacht racing and can be dangerous if the yacht tacks unexpectedly so that the water ballast is suddenly on the leeward side. Tabarly won the race covering the 5700 miles in 40 days and arriving at Yokohama on 29 April 1969 (5·9 knots).

This passage of Tabarly is the fastest Pacific crossing in a single-hander, though comparisons are difficult because there are many varied routes to different shores in the Pacific. The fastest passage made in the Pacific which may also be the fastest ocean run between fixed landmarks made by any sailing yacht is by *Pen Duick IV*, the trimaran, when fully crewed. She sailed a multihull race between Los Angeles, USA and Honolulu, Hawaii, 2225 miles in 8 days 13 hours which is 260½ miles per day average or 10·854 knots. The fastest Atlantic crossing by a yacht is also held by *Pen Duick IV*. In 1968, fully crewed, she sailed from the Canaries to Martinique, covering the 2640 miles in 10 days 12 hours, that is on average 251·4 miles per day (10·47 knots). Comparable times under sail have only been achieved by far larger yachts from west to east in the North Atlantic, the records being the schooner *Atlantic*, Sandy Hook, New York to the Needles, Isle of Wight in 1904 taking 12 days 4 hours. The four-masted commercial sailing ship *Lancing* made the passage from New York to Cape Wrath in 1916 in 6 days 18 hours, an average of 14·8 knots. But this wartime gale-wracked run was quite different in purpose, crew size, ship, and circumstance.

Round the world

After the completion by Francis Chichester of his voyage in *Gypsy Moth IV* in the spring of 1967, some single-handed sailors were bound to think about further endeavour. Almost any feat seemed possible, although it must be remembered that Chichester only just made Sydney in a state of exhaustion. However he had shown the way and he was older than would be expected. By the end of 1967, each of four yachtsmen was preparing to sail round the world *non-stop*. Each had experience of crossing oceans before. The first to announce his intention was Commander W. D. 'Bill' King (Brit.) who had a special 42 ft hull designed by Angus Primrose and built by Souter of Cowes. This had a Chinese junk ketch designed by Blondie Hasler. Second

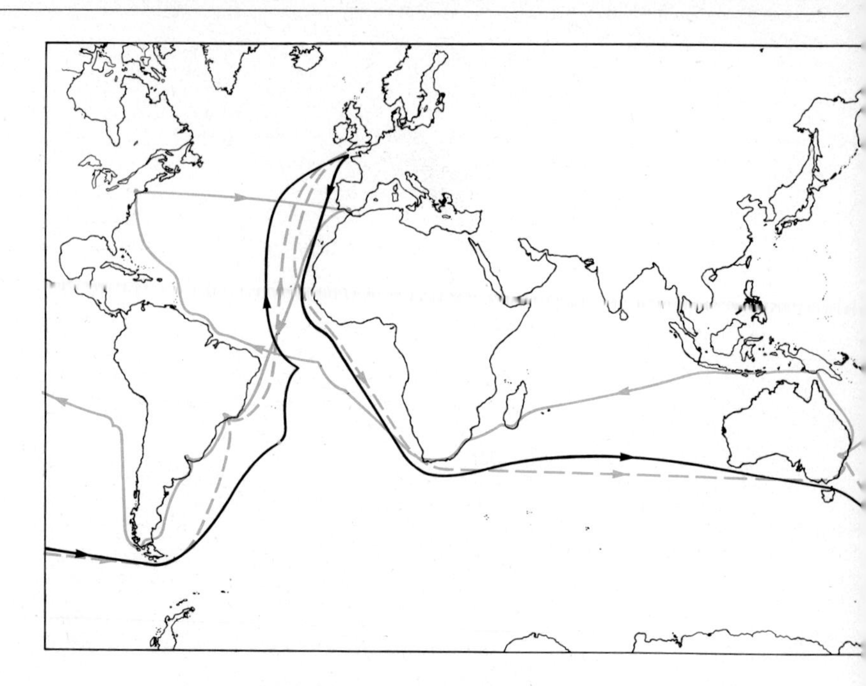

there was Robin Knox-Johnson (Brit.) with *Suhaili* (see p. 156). Thirdly, there was Bernard Moitessier (Fr.) in *Joshua* (see p. 162). Fourthly, John Ridgway (Brit.), who had rowed the Atlantic with Chay Blyth, was equipping a 30 ft glass-fibre bilge-keel sloop.

The first and only single-handed race round the world was declared by the London newspaper *Sunday Times* in March 1968. No formal entry was necessary, so those already preparing to sail were in a race whether they wished to be or not. Several further sailors decided to attempt the voyage and a total of 8 started out. There was no formal start: instead there were two main awards. These were the 'Golden Globe' for the first yacht home and a £5000 prize for the yacht to make the fastest voyage. It was necessary to start from any port in the British Isles, any time between 1 June and 31 October 1968. Return had to be to a port in the British Isles. The idea was literally disastrous as mentioned before. This was partly because of the short time from the announcement of the 'race' until the last date on which the competitors could leave.

Victress, the trimaran which encircled the world, but broke up before reaching England.

The only yacht to complete the race was *Suhaili*, sailed by Robin Knox-Johnston. He became the first man to sail non-stop round the world and, therefore, won both trophies. Two other competitors did make round the world trips, but did not fulfil the conditions of the race. Moitessier sailed on to Tahiti. Nigel Tetley (Brit.-S.A.) sailing the 40 ft trimaran *Victress*, a ketch with 900 ft² sail area and beam 22 ft, left Plymouth on 16 September 1968 and passed by the Cape of Good Hope, Australia and New Zealand and Cape Horn: only Knox-Johnston and Moitessier achieved this in the race. Tetley then crossed his outward track in mid-Atlantic in the region of the Equator on 22 April 1969. In this respect he circled the world faster than any other non-stop single-handed sailor, by completing this passage in 179 days. However the voyage was not completed because on 21 May his trimaran began to break up quickly and he was forced to take to his life-raft when some 800 miles west of the Spanish coast. By means of a radio distress signal he was spotted by aircraft of the US Air Force based in the

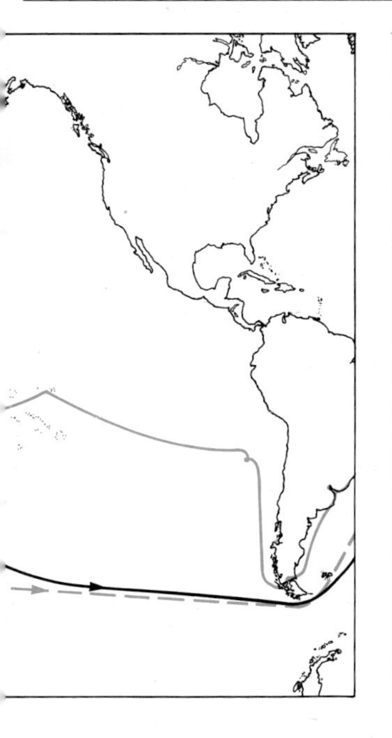

Voyaging round the world. Joshua Slocum sailed from Boston to Gibraltar and then west about route coloured solid blue. He called along the South American coast, went through the Strait of Magellan, across the Pacific in the South-East trade winds to Sydney. From there he took the trade winds through the Indian Ocean and across the South Atlantic before heading for Central America and back to Boston.

Robin Knox-Johnston, the first man to sail alone round the world without any port of call left Falmouth, England, went down the South Atlantic. Then most of his route was in the Roaring Forties and strong westerlies of the Southern Hemisphere, from which he rounded Cape Horn and sailed up the Atlantic and back to England. Route coloured black.

First fully crewed Round the World race, 1973–4, had prearranged ports of call. From Portsmouth, the yachts went to Cape Town, Sydney and Rio before returning to England. Much of the route (broken blue line) was in the Roaring Forties including a rounding of Cape Horn.

Azores and picked up the same day by the tanker *Pampero* without injury.

The greatest hoax on a single-handed voyage, but a tragic one, was by Donald Crowhurst (Brit.). He left on the Round the World race from Teignmouth, England, on the last eligible date and sailed for the South Atlantic. He also sailed a 40 ft Piver-designed trimaran ketch, *Teignmouth Electron*. Crowhurst sent back radio signals during his voyage and his position was plotted by the *Sunday Times* as he crossed the southern ocean, rounded Cape Horn and then headed back up the Atlantic in the early summer of 1969. It was because his position looked very good that Tetley pressed his own trimaran which may have been the cause of it finally breaking up. However, *Teignmouth Electron* was found by a ship in mid-Atlantic, adrift in good condition but without Crowhurst on board. This was on 10 July and the last log entry was 23 June. It was later found by examination of the log that Crowhurst had never left the Atlantic. He had sailed into the South Atlantic, in fact a total of some 8000 miles of ocean sailing, had landed briefly on a remote part of the Argentinian coast and had feigned signals stating he was in the Roaring Forties. Unfortunately the entries in the log showed that he was a schizophrenic and it was presumed that he had flung himself overboard. In another tragedy Nigel Tetley committed suicide in 1972.

None of the other competitors in the race got as far as rounding the Cape of Good Hope. John Ridgway departed in *English Rose IV* from the Arran Islands (where the rowboat *English Rose III* had previously made a landfall), but was forced into a South American port with damage, and could not continue. Chay Blyth who had decided to enter the race in a standard glass-fibre Kingfisher, 30 ft, *Dytiscus III*, put into Tristan da Cunha and finally into East London, South Africa, with broken self-steering gear and other damage. He was joined there by his wife and sailed back two-handed to the Azores. The Frenchman Louis Fougeron in the gaff cutter *Captain Browne* was damaged in the South Atlantic and put into St Helena. The Italian Alex Carozzo, in the biggest yacht to enter the competition, the

Vendredi Treize: largest single-handed yacht

Penduick VI starts on her race round the world, skippered by Eric Tabarly

Sayula, coming in towards the finish as she wins the first ocean race round the world

Chay Blyth in *British Steel* arrives home after his unique voyage

66 ft ketch *Gancia Americano*, was taken ill at an early stage and returned to Europe. The original adventurer, Bill King, with his specially designed craft, was capsized by a huge wave 800 miles west of Cape Town and his Chinese rig swept away. He managed to set up a special emergency rig which had been planned and reach Cape Town. Over the next few years which included further dismasting and damage, he sailed round the world and back to Britain.

Single-handed races are generally unusual, though they may be run by yacht clubs over short courses as a 'novelty' event. For any longer courses, a major problem is to keep the yachts clear of shipping (for which they are not looking out). Up to 1974 there had, however, been five repeats of such a race organized by the French newspaper *L'Aurora*. This 'course solitaire' had proved successful and with the course in the general area of the Bay of Biscay. The actual course has been amended a few times, but typical is a start from Perros-Guirec (north Brittany), then to Falmouth, England, with short compulsory stop; to Kinsale, Ireland; to Laredo, Spain; to finish at Pornic (near the entrance of the Loire River). Winners have included Jean de Kat, M. Malinovski, Jean-Marie Vidal, and Gilles le Baud.

Single-handed races were sailed in 1975 from Falmouth to the Azores, and back for boats not exceeding 38 ft.

Fully crewed events

The first fully crewed ocean race on a course round the world began on 8 September 1973. The organization of this event was clearly a reaction to the ill-advised aspects of the *Sunday Times* single-handed race. For instance, it was organized by the Royal Naval Sailing Association with aid of worldwide naval communications; all yachts had to be properly crewed with a minimum of five persons; the International Offshore Rule was to be used with a handicapping time-on-distance scale so that the winner would not necessarily be the largest yacht if that was first boat home; there were to be compulsory stops and co-ordinated fresh starts at Cape Town, Sydney and Rio de Janeiro, the race having to begin and end at Portsmouth, England; there was a minimum size of 33 ft rating (about LOA 45 ft); no multihulls were allowed. Nevertheless the course was still south of the three great capes of the southern ocean – Good Hope, Leeuwin, and the Horn.

Sayula II, Swan 65 class designed by Sparkman and Stephens and winner of the first Round the World race for fully crewed ocean-racing yachts.

The 77 ft 2 in *Great Britain II*, first to finish in the Round the World race in 1973–4.

This was the longest-ever race for ocean-racing yachts covering the following nominal distances:

Portsmouth to Cape Town	6650 miles
Cape Town to Sydney	6600 miles
Sydney to Rio de Janeiro	8370 miles
Rio de Janeiro to Portsmouth	5500 miles
The total for the race was	27 120 miles

The winner of this first Round the World race was the 64 ft ketch *Sayula II*, owned and sailed by Ramon Carlin (Mex.). She was a production Swan 65-class yacht in GRP designed by Sparkman and Stephens of New York and built by Nautor Ky, Finland. Her placing in each respective leg – 2nd, 1st, 2nd, 4th. These positions gave her the best aggregate corrected time for the entire course. *Sayula*'s crew consisted of 12 on any one leg and included the sons of the owner and Americans, Britons, and Mexicans. Her actual time was 152 days 9 hours (average speed 7·41 knots).

The boat to make the best elapsed time was the ketch *Great Britain II*, sailed by Chay Blyth and men from the Parachute Regiment, who had not before been sailing men. The yacht, designed by Alan Gurney (USA), was of foam sandwich and built in England – LOA 77 ft 2 in, LWL 68 ft 2 in, beam 18 ft 5 in, draught 9 ft, displacement 73 000 lb (the IOR rating was the largest allowed at 70 ft). Her sailing time was 144 days 10 hours (7·82 knots), which is the record time for any yacht circumnavigating the world.

The most disappointed man in the race was Eric Tabarly, skipper of the aluminium ketch *Pen Duick VI*. His 74 ft yacht was specially built by the French Navy at Lorient and designed by André Mauric, a foremost French designer. On the first leg the mainmast broke and Tabarly took his boat to Rio, to where a new spar was flown out by the French Navy.

33 Export, one of the French entries, completes the Round the World race. The skipper had been lost in the Roaring Forties.

He sailed on to Cape Town, then completed the 2nd leg in which *Pen Duick VI* was first to finish and placed 5th. On the 3rd leg the mast again broke and the yacht put back to Sydney to have it repaired. The 3rd leg was completed, but Tabarly abandoned the 4th leg and sailed straight on to Brest.

The largest yacht to start was the 80 ft ketch *Burton Cutter*, owned by Leslie Williams and Alan Smith and built of aluminium in England. She gave up after suffering hull damage early in the 2nd leg after leaving Cape Town. She sailed back across the Atlantic and joined in the last leg. Altogether 17 yachts started and 14 did every leg of the course. The participants were by nationality – British: *Adventure* (54 ft 6 in), *British Soldier* (ex-*British Steel*) (59 ft), *Burton Cutter* (80 ft), *Great Britain II* (77 ft 2 in), *Second Life* (71 ft); French: *Grand Louis* (61 ft), *Kriter* (68 ft), *Pen Duick VI* (74 ft), *33 Export* (57 ft); Italian: *CSeRB* (50 ft), *Tauranga* (55 ft), *Guia* (45 ft); Polish: *Copernicus* (50 ft), *Otago* (55 ft); German: *Peter von Danzig* (59 ft); Mexican: *Sayula* (64 ft). The following did one leg: *Jakaranda* (S.A.), 1st leg; *Concorde* (Fr.), 2nd leg; *Pen Duick III* (not *VI*) (Fr.), 4th leg.

No less than 14 of the competing yachts had been built since 1970 and several specially for the race. The oldest boat to compete was the 1936-built *Peter von Danzig* sailed by young Germans. Boats built of steel, aluminium, wood, foam sandwich and GRP all completed the course.

Ramon Carlin, owner and skipper of *Sayula II* with his wife at the end of the race.

Several women completed one or all legs. Wendy Hinds (Brit.) on *Second Life* went the whole way: Yvone van de Byl sailed on *Sayula* on the 2nd leg during which the yacht received a knock down which put the masthead below the water; Mrs Carlin sailed on the 1st leg. Mrs Pascoli wife of Eric Pascoli sailed in *Tauranga* which he skippered. Miss Christina Monti – at 18 the youngest crew – and Mrs Carla Malingri (wife of skipper) sailed in *CSeRB*.

The most disastrous occurrences during the race were the loss of three men, all on different occasions. The skipper of *33 Export*, Dominic Guillet, was lost on the 2nd leg when a sea broke his lifeline and swept him away. From *Tauranga*, Corporal Paul Waterhouse, who had joined her from *British Soldier* for the 2nd leg was also washed overboard. From *Great Britain II*, Bernard Hosking was knocked overboard when working on the foredeck on the 3rd leg in the cold and gales of the Roaring Forties and was not recovered despite searches by Chay Blyth and his crew. In the rules of ordinary yacht racing, if a man is lost overboard the yacht must recover him or she is not a valid finisher, but there was a clause in the Round the World race allowing persons to leave. It was not originally intended for losing crew overboard, but it became an escape clause for the yachts to continue. It was argued that they had to go on to the next port anyway: there is no doubt morale would have been even worse, after losing a man had they sailed there without being able to race for it. No other yacht race has suffered comparable incidents, but then no other race has covered such sailing distance. The time spent sailing by the yachts was equivalent to many years of the sailing life of a normal week-end and holiday ocean racing boat.

The Round the World race although organized by the Royal Naval Sailing Association – and with the prizes finally presented in London by Prince Philip – was sponsored by Whitbread and Co. Ltd. *The Financial Times* announced another race to start from London in August 1975. The course was to be the same as the previous race, but with a stop only at Sydney. Eligible were fully crewed yachts of up to 70 ft rating but with no time allowance for the main prize. Therefore the most serious competitors would have to be at or close to this rating, i.e. yachts of about LOA 76 to 85 ft. Considerably more stores would have to be carried than in the shorter legs of the other race, or compared with Sir Francis Chichester who did the whole distance, but only had to feed one! One of the intentions of the organizers was that a yacht in the race should beat the 144 days of *Great Britain II* and the 136 days of the clipper ship *Patriarch* over the same routes. Whitbread were arranging another race for 1975 starting on 1 November – an ocean event of 16 000 miles purely for fully manned multihulls from Portsmouth to Key West, Florida, via the Azores, and with further legs criss-crossing the Atlantic. These were Key West via Barbados to Freetown, Freetown to Rio de Janeiro, Rio via Cape Verde Islands to England. A race round the world for multihulls in the mid-1970s did not seem suitable for their stage of development and race and rating situation. Round the world races for both types of hull were scheduled for 1977 and 1981.

ROWING AND SURVIVAL

Why row across an ocean? Sailing may be rugged and slow, but sails are still the most logical method of propulsion for small craft, being more dependable than small engines, for which it may also be very difficult to obtain enough fuel in out of the way spots, or for which it becomes impossible when a certain small size is reached to find stowage and space for fuel. But rowing? It is a sort of sailing anyway, for deliberate rowing expeditions fight the wind and still expect to reach their destinations. So they must sail with the prevailing wind (and current) yet deliberately forego spreading any area of fabric so that the wind can speed them. They drift with the elements and add a minute propulsion forward or to one side or the other to pilot themselves – yet with appalling human effort. To add to the hardship a rower must for much of the time be exposed to the weather for much of his body has to be conspicuous to row at all. A sailor can keep below in inclement weather except when attending to gear on deck. And nowadays his steering vane and simply handled sail plan can be left for many hours while he makes progress, checking his compass below decks and his other instruments over the chart table. Only in a windless calm and flat sea would the rower be able to smile at the impotent sailing boat.

When asked why he had performed such a feat, one man who had rowed across the Atlantic said that, among other reasons, it was to belong to a very select club indeed. Another, John Fairfax, said, 'Personally I have sworn never to touch an oar again and you will not see me on the Serpentine for a million dollars.' Very few men have ever made this 'unnecessary journey'; some others have been forced to because they have taken to boat or raft after a ship has sunk. A few have set out with the intention of rowing and have given up or been lost at sea.

The first known two-man Atlantic crossing by oars alone in the *Richard K. Fox*, 18 ft.

The first men to row intentionally across an ocean were George Harvo (Norw.), aged 31, and Frank Samuelson (USA), aged 36 in 1897. In their double-ended (pointed stern) clinker-built 18 ft open boat, *Richard K. Fox*, with buoyancy tanks and spare oars they took 55 days from New York to St Mary's, Scilly Isles, off south-west England. This 3075 mile journey took from 6 June to 1 August 1897. This remains the fastest east to west crossing with oars. During the voyage they were swamped and turned over, but righted the boat and bailed her out. From the Scilly Isles, Harvo and Samuelson rowed to Le Havre and *Richard K. Fox* was exhibited in Paris. It is thought that the rowers, who were sailors – New Jersey oyster fishermen – may have been sponsored by a Mr Richard K. Fox.

There were no further ideas of rowing across the Atlantic – or any other ocean – until 1965 when David Johnstone (Brit.) decided to row from the USA to the British Isles and commissioned a boat, specially designed by Colin Mudie, to be built by a Cowes boatbuilder. Colin Mudie besides being a top yacht designer had sailed across the Atlantic in *Sopranino*, 21 ft, in 1951. He had also landed in mid-Atlantic in a balloon, *The Small World*, the gondola of which was a boat of his own design. Leaving Medano in the Canary Islands in December 1958 he had remained airborne for 1200 miles. After ditching, the semi-catamaran hull with its crew of four travelled the 1500 miles to Barbados, but it did have mast and sails. The Johnstone boat was called *Puffin*. She was 15 ft 6 in long with a beam of 5 ft 6 in and was partly decked with hatches, buoyancy, rudder, and radar reflector. There were hand-holds on deck and on the bottom in case of capsize.

During the preparations, Captain John Ridgway of the Parachute Regiment had discussed whether he should be the second man in *Puffin*. However, Johnstone took John Hoare, a journalist. Ridgway decided to make his own expedition enlisting Sergeant Chay Blyth, another parachutist, as his companion. A kind of race therefore began.

Puffin left Virginia Beach, Norfolk, Va. on 21 May 1966. After 13 days it was only 120 miles from the American coast. *Puffin* was found on 14 October awash without her crew 800 miles east of St John's Newfoundland, by a Canadian warship: she is thought to have been overwhelmed by Hurricane Faith early in September, 100 days out and still with more than 1000 miles to go to the British Isles.

Tom Maclean only single-handed Atlantic rower from west to east.

The first men to row across the Atlantic in this century were John Ridgway and Chay Blyth (Brit.). Their boat, *English Rose III*, was a Yorkshire dory (20 ft, beam 5 ft 4 in), an open boat to which was added turtle decks fore and aft over buoyancy compartments. The freeboard was increased and the gunwhale strengthened. Four pairs of oars were carried to give ample spares. *English Rose* was a standard wooden design built by Bradford Boat Services in Yorkshire. Chay Blyth had no experience of the sea, but both men, as parachute troops, were fit and strong with the right mental attitude for adverse conditions. They chose to leave from Orleans, Cape Cod, Mass. on 4 June 1966, but met more adverse winds than expected. At first the passage was slow, but once in the Gulf Stream current, *English Rose* made better progress. On 13 August in mid-Atlantic they went aboard the tanker *Hanstellum*. Landfall was made at Inishmore, Aran Isles, Ireland on 3 September, after a crossing of 91 days.

The next successful attempts to row across an ocean were single-handed ones across the Atlantic. In 1969, John Fairfax (Brit.) had the 22 ft *Britannia I* designed by Uffa Fox. He was the first man ever to row across an ocean alone and left Port St Augustin, Grand Canary, Canary Islands (to which the boat had been brought by steamer on 20 January 1969) and arrived at Fort Lauderdale, Florida, on 19 July without putting in anywhere. This was the slowest and longest of all Atlantic rows, taking 180 days.

Super Silver, Maclean's boat, the Yorkshire dory of the same design as *English Rose III* which is now in Exeter Maritime Museum.

The only man to have rowed across the Atlantic from west to east by himself was Tom Maclean (Brit.) who also crossed in 1969. He rowed from St John's, Newfoundland, and arrived in Blacksod Bay, Ireland, in 70 days. His boat was a 20 ft Yorkshire dory (the same as that used by Ridgway, but of course with only a single rower). His route closely resembled that of *English Rose III*, but the starting-point being even farther north no doubt helped him despite having half the oar power.

The longest solo rowing voyage was completed in 1970 by Sydney Genders (Brit.), aged 51, who rowed alone across the Atlantic first from Sennen Cove, England to the Canaries, continuing from the Canaries to Antigua and then going on from there to Miami Beach. In his 20 ft boat, *Khaggavisana*, his was also the fastest solo row from east to west. He is the oldest person to make such a journey.

The fastest passage made from east to west across the Atlantic by rowing was that of Geoff and Don Allum (Brit.), cousins aged 23 and 33, who also chose to use a 20 ft Yorkshire dory of the same design and by the same builder as *English Rose* and the Maclean boat. The dory was taken by steamer to Las Palmas and the two men rowed in 74 days to Bridgtown, Barbados, West Indies.

The only deliberate crossing of the Pacific by a rowing-boat was made by John Fairfax and Sylvia Cook in 1971. Miss Cook is the only woman to have ever engaged in a deliberate long ocean row. The boat used, called *Britannia II*, was developed from the experience of Fairfax after his 1969 row. She was again designed by Uffa Fox and built by Clare Lallow of Cowes of double-planked Honduras mahogany, much longer than any boat used by the other rowers, she was 35 ft, beam 5 ft 2 in, draught 1 ft. There were shelters fore and aft topped with self-righting blisters made of plastic foam. The larger boat meant that a considerable amount of gear could be carried including five pairs of oars, two sextants, range-finder, two chronometers, two transistor radios, a transceiver Marconi radio, camera equipment, spearguns, shark-shooting equipment, distress beacons, and all the food and 90 gallons of drinking-water that would be required for traversing such a long distance in a hot climate. *Britannia II* left San Francisco on 16 April 1971, but the weather forced them back down the coast to Ensenada, Mexico, where the pair arrived on 3 June. They left there on 17 June but four days later found their rudder had disappeared and they accepted a tow back there. Fairfax and Cook left again on 26 June. This was already the 62nd day of the trip from San Francisco. On 6 October the boat arrived at Washington Island in mid-Pacific and they stayed there until 12 November when they rowed on. On 9 January in the Gilbert Islands group, they were driven ashore in breaking seas on an island called Omotoa. The boat was damaged, but local people hauled her out and she was repaired in the local boatyard on another island, Tarawa. The rowers left there on 7 February 1972 heading for Australia and in this stretch Fairfax was badly bitten by a shark when trying to spear other fish. As a result they tried to attract the attention of ships with an alarm beacon and visual means, but two ships sailed past close without seeing them (this is unfortunately the common experience of people in small boats trying to attract attention). For four days they were hit by cyclone Emily: 'We lay together in the rat hole while *Britannia* weathered the storm . . . crashing and bumping but always rising at the right moment. Time and again we watched a 20 or 30 foot wave break only yards fore or aft, but never over us.' The boat reached Hayman Island on 22 April 1972, four days under the year which they first set out from San Francisco. At times it was widely assumed that they must have been lost on the longest-ever rowing voyage of 9000 miles. John Fairfax's total rowing mileage was 12 000 miles, by far the greatest of anyone.

Maurice Bailey boards the inflatable dinghy on which he and his wife spent 118 days drifting in the Pacific. In the picture their yacht *Auralyn* is sinking and the stress is reflected in Bailey's face. His wife had the presence of mind to take this photograph in a critical situation.

The same ocean rowing-boat, *Britannia II*, was borrowed by Derek King and Peter Bird (Brit.) for an attempt to row round the world, together with a girl, in 1974. Originally scheduled to leave Gibraltar, they had what appears to be a common difficulty for rowers, of getting away from a coast, and eventually left Casablanca without the girl companion. They reached St Lucia, West Indies, 93 days later after a gruelling trip in which they were nearly run down by two ships. On arrival it was stated that they were mentally exhausted and the idea of the circumnavigation was abandoned.

Involuntary ocean voyages are more common, mainly due to shipwreck and they include for instance the historic voyage of Captain Bligh in the boat of the *Bounty* in 1789 from off Tahiti to Timor. This was a distance of just

Q.E.3, yet another Yorkshire dory, was used by Geoff and Don Allum who made the fastest passage from east to west under oars alone.

under 4000 miles, 3 months' rowing and 18 men, some of whom died, in a 25 ft boat.

In 1923 the SS *Trevessa*, 1200 miles off the western coast of Australia, was lost and two of her boats rowed over 1000 miles to reach Rodriguez Island and Mauritius. These were, however, proper ship's lifeboats with oars and sails and manned by professional sailors with senior ship's officers in charge of them. Other such voyages have been always a very great feat.

The longest recorded survival by anyone is that of Second Steward Poon Lim in a ship's lifeboat in the Atlantic for 133 days. The boat had carried supplies and water which he was able to consume. His ship, the SS *Ben Lomond*, had been torpedoed in the Atlantic 750 miles off the Azores on 23 November 1942. He was picked up on 5 April 1943 and was later awarded the BEM.

The longest survival in peacetime and the longest in inflatable rafts or dinghies is by Maurice and Maralyn Bailey, whose 28 ft yacht *Auralyn* was sunk by sperm whales north-east of the Galapagos Islands in the Pacific on 4 March 1973. They took to a four-man life-raft and 9 foot rubber dinghy which were tied together and proceeded to drift for 118 days until rescued by the Korean fishing-vessel *Weolmi 306* on 30 June. They existed on solids and liquids obtained from fish, turtles, and birds after using up the small amount of supplies in the dinghies. Drinking-water was obtained by catching rain. During the period in which they were adrift no less than seven ships were sighted, some as close as half a mile. Despite flares and waving and calm sea conditions on occasions, none of these ships appeared to see them. It was concluded that in every case a proper lookout was not being kept and that as the ship was in a remote part of the ocean, where it was unlikely to meet others, that it was set to automatic pilot. One ship did stop and made two complete turns, but then steamed away apparently not having seen the castaways or having mistaken them for some other object in the sea.

In an almost parallel circumstance in nearby waters, 180 miles west of the Galapagos, the schooner *Lucette* was sunk on 15 June 1972, also by being attacked by whales. On board were Dougal Robertson, his wife, 18-year-old son and 12-year-old twins, and a student friend. They took to an 8 ft solid dinghy and life-raft in poor repair, which sank on the 17th day adrift. After 38 days they were picked up by the Japanese fishing-vessel *Tokamaru II*. They also sighted ships and sent off flares and other signals, but as in the cases mentioned above, there was no reaction and the vessels steamed onward. They existed off fishes and turtles and birds and all the party survived.

5 Design, Building and Equipment

DESIGN OF YACHTS

Edward Burgess

The earliest method of designing a boat or yacht was by carving a model until it looked the right shape. One might take a model of a hull and make the bows flatter or the keel longer. In yachts the same procedure is followed today in the sense that 'a bit more here, or a bit less there' is practised. The use of lines plans, mathematics, and computers speeds up and clarifies matters, but yacht design is still basically an art rather than a science. Models are still used, but for tank testing.

The earliest designs for yachting (talking now of the post-war period beginning in 1815) were those of warships and workboats. If there was a reputedly fast fishing-boat, then it was reproduced for private use. There was, therefore, little individuality of form, but like the fast smugglers, there was a round barrel-like bottom and a flat run aft to the hull. All the vessels were built of timber and ballasted with stones or gravel inside. Main rigging was of hemp, sails were baggy and of flax.

A typical boat, but remarkably successful in racing, was *Arrow*, a cutter built in 1823 whose dimensions were LOA 61 ft 9½ in, beam 18 ft 5¼ in, depth of hold 8 ft 8 in.

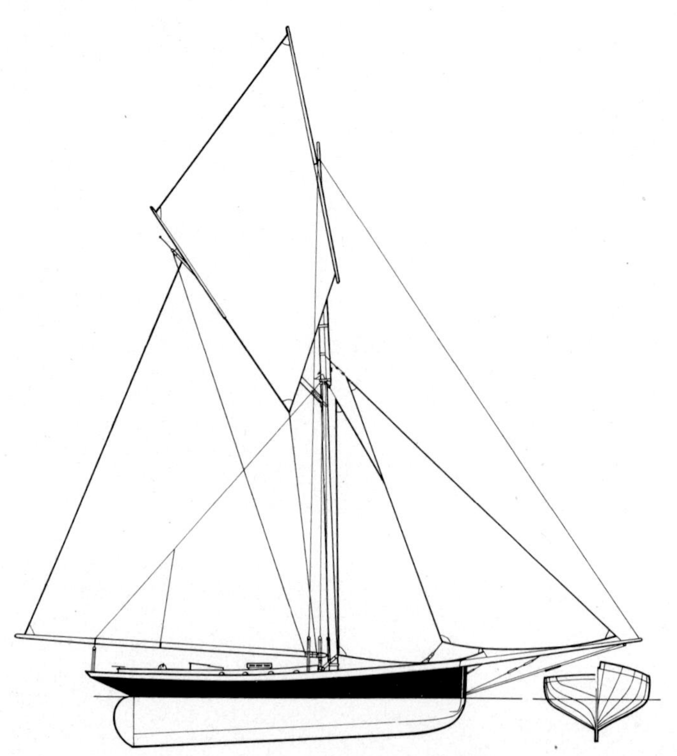

The 69 ft 9 in *Vindex* was the first yacht to be designed entirely on paper, rather than using a model. The designer was A. Carey-Smith of New York and the yacht was launched in 1870.

Jullanar, 1875, was in her time a breakthrough in yacht design.

Right
Gloriana, designed by Herreshoff in 1890.

Designers

The first use of the term 'yacht designer' was not until long after individual yachts were already being designed. The year 1884 was the first in which a distinction was made in *Lloyd's Register* between builders and designers. The first person to use the term may have been Dixon Kemp (born 1839), author, designer, and first Secretary of the Yacht Racing Association. Kemp reported that St Clair Byrne, 'started as a naval architect making yachts a speciality in 1868; just when the rage for 10-tonners began to be manifested'. George L. Watson of Glasgow began designing in 1871. The design firm of G. L. Watson is still active today at Erskine, Renfrewshire.

The first American to be considered primarily a yacht designer was A. Cary-Smith (born 1837 in New York). In 1870 with Robert Center of the New York Yacht Club he designed the first yacht to be completely drawn and specified on paper (in the modern way) before building. Based on the successful British cutter of the time, *Mosquito*, the yacht was of iron. The bowsprit, also of iron, was half the length of the cutter and weighed 800 lb! She was the first iron yacht in America. The yacht called *Vindex* was length 69 ft 9 in, beam 15 ft 3 in, draught 10 ft 6 in, sail area 5700 ft^2.

Smith was one of the earliest official yacht measurers, being Club Measurer of the New York Yacht Club from 1872 to 1882. He designed the America's Cup defender *Mischief* in 1879 and hundreds of yachts and also river passenger steamers.

The most distinguished American designer of the later nineteenth century was Edward Burgess (born 1848 in Boston). He spent a summer visiting the south coast of England and later made adaptations in the USA of the traditional Itchen Ferry boats. After designing the successful America's Cup defender *Puritan* in 1885, orders for numerous designs poured into Burgess's rapidly expanded office. In seven years his office designed 137 different yachts; one task was modernizing the original *America*. His son Starling Burgess became equally famous as a yacht designer. One of Starling's last yachts was the 1934 America's Cup defender *Rainbow*, in which the young Olin Stephens co-operated. Stephens could be said to have carried forward the banner of American yacht design, which has invariably led the rest of the world.

Carina, owned by Richard Nye, 1969. She represents the fine type of yacht designed to the Cruising Club of America Rule, which was superseded by the IOR within two years

On board the British Admiral's Cup contender *Marionnette* in 1975. It shows typical modern gear, the powerful winch holding the spinnaker guy, tracks for fairleads, other winches and instrumentation

Victorian design

The first sailing yacht in Britain to break away from working-boat form was the 110 ft 6 in yawl *Jullanar* of 1875. Her designer, E. H. Bentall, was an agricultural machinery maker by trade in Essex and he sailed on the Blackwater River. *Jullanar* was designed by modelling, but the key to her success was the cutting away of profile and reduction of wetted surface (i.e. water friction). Other hull dimensions were LWL 99 ft, beam 16 ft 10 in, draught 8 ft 6 in. The beam was narrow because of current rating rules. The rounded keel and canoe stern were unconventional at the time when plumb stems, straight keels, and long counters were prevalent.

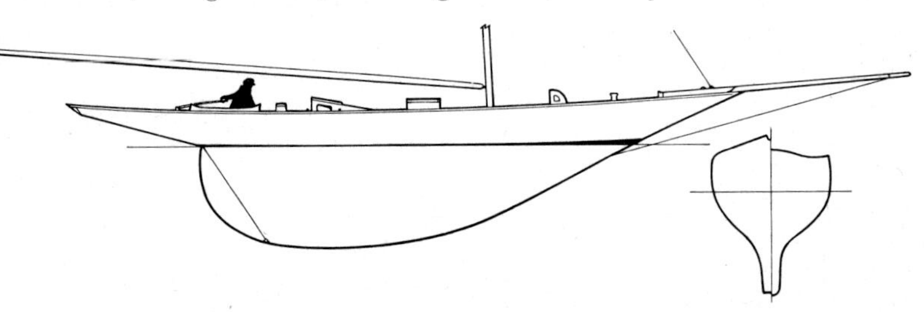

In *Gloriana*, Herreshoff abolished straight stems, cut away the profile and turned the long keel into a real fin (note section).

The first yacht to have a distinct fin keel with ballast on it was designed by Bentall in 1880. Called *Evolution* she was not a success because of the narrowness (6 ft 6 in beam on a 50 ft 9 in LWL!) caused by the rule then in force. In 1887 G. L. Watson designed *Thistle*, with the cutaway hull type of *Jullanar*, to challenge for the America's Cup but the yacht did not have enough distinct 'fin' to her keel and made too much leeway. This was not a fault in *Gloriana* designed by Nathanael Herreshoff (US) in 1890 with LOA 70 ft, LWL 45 ft 4 in, beam 12 ft 7 in, draught 10 ft, sail area 4137 ft^2. She was a great success in the '46 ft class' to which she belonged and to the length and sail area rule (see p. 190). She had a fin, cutaway keel, and new overhang bow profile and sailing yacht design, at least where some speed was wanted, never returned to the old forms. Three years later G. L. Watson designed *Britannia* to the same ideas and with great elegance for the Prince of Wales. This 1893 design came to be considered an ideal form for 40 years afterwards. She was subsequently raced successfully to three different rating rules.

The most disastrous failures in yacht design were caused by extremes. On the British side the development of ever-narrower beam (owing to the rating rules then in force) caused William Evans Paton to design in 1886 the cutter *Oona* with a beam of only 5 ft 6 in on an LWL of 34 ft. Draught was 8 ft, displacement 28 000 lb and ballast 21 504 lb – a very high percentage of her weight was in lead. She was, however, built very meticulously.

On her first voyage from the Solent to the Clyde, the yacht was driven ashore on the Irish coast and the heavy keel torn from the light hull. All hands, including the designer who was on board, were drowned.

The comparable American disaster was due to the same basic cause: extremes in design development. The schooner *Mohawk* was built to 'lick all creation' in 1875 with LOA 150 ft, LWL 121 ft, beam 30 ft 4 in. Her main boom was 90 ft, her foreboom was 39 ft, but her draught was only 6 ft, though 31 ft when the long centreplate was lowered. So whereas *Oona* was a 'plank-on-edge' (the current epithet and still used for narrow boats), *Mohawk* was a 'centreboard bug'. The owner of *Mohawk* was the newsprint millionaire William T. Garner, the builder and modeller was Joseph Van Densen of Williamsburgh. On 20 July 1876 as some sails were

The 150 ft American schooner *Mohawk*, built to 'lick all creation' had only 6 ft draught, though a 31 ft centreplate could be lowered. She represented the extremes of yacht design reached in 1875. Within one year of building she had capsized with loss of life.

The sections of *Jullanar* 110 ft 6 in.

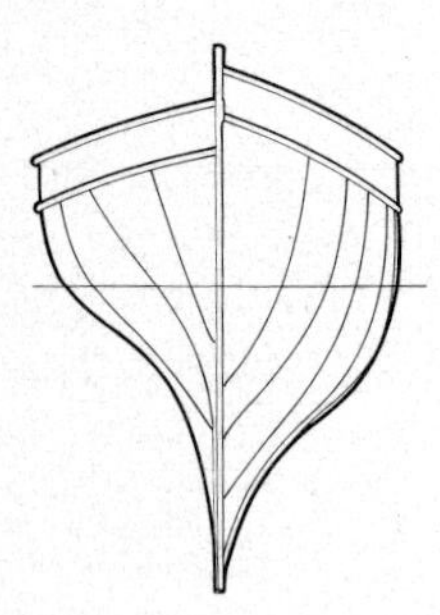

being hoisted and the crew was weighing anchor the yacht was heeled by a puff of wind off Staten Island and capsized. The owner and his guests, who were below, were drowned.

Several other capsizes of American yachts without adequate draught and low ballast led to a veer away from such an extreme type. And the result of the loss of *Oona* was contributary to a change in the rating rule under which she was designed.

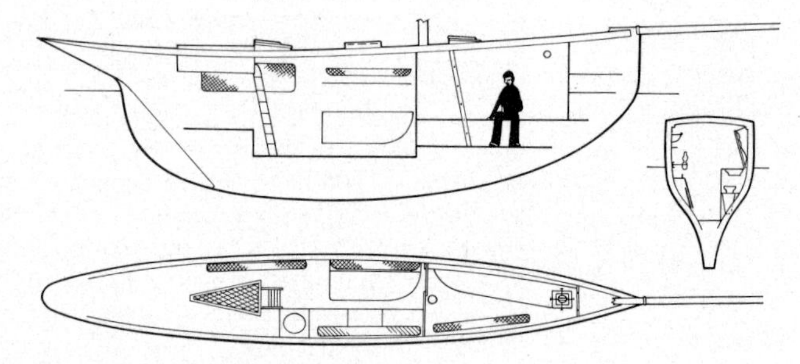

Extreme design: the plank-on-edge cutter *Spankadillo* 3-tonner, designed and built in 1882. It was the product of a rating rule which taxed beam, but little else.

Rules of rating and measurement

The greatest influence on the design of sailing yachts is the racing of them. Even if intended purely for cruising, many features of great importance in the hull and rig are derived from racing practice. This may not be current racing practice, but it will have come from the attempt of designers to make boats go faster in previous years. The common Bermuda rig itself is an example, as are lead ballast on a keel, lifelines, winches, and special propellers. Such effects as chopped-off counter sterns, which look 'modern', come from an effort to get an advantage from a rating rule. The greatest single influence outside the natural ones of weather, the sea, ergonomics, and the constraints of current material technology is the rule of measurement and rating in force. The history of these rules is to a great extent the story of sailing yacht design, as yachtsmen have striven to win under the self-imposed restrictions of these rules. They are not quite so artificial as might be supposed because commercial sailing-ships were trying to do the same thing against a pay-load of freight, warships against the maximum armament, pilot-boats against minimum manning for long waiting periods (compare this with the two-man Round Britain and such short-handed races).

The first attempts at rating yachts were, therefore, by using the existing tonnage rules. A 'tun' was a barrel and from medieval times vessels had been measured for taxation by the number of such tuns that could be carried – a just arrangement. But the ancient owners – like today's rule-avoiders – did their best to squeeze advantage from such a law. In order to save actually counting the 'tuns' there was a simple measurement: by Act of Parliament in 1694 it was

$$\frac{L \times B \times D}{94}$$

L was the length along the keel, B was the breadth, and D the depth of hold. Directions were given to measurers so that the ships should be measured upright and the keel length properly found. As depth was difficult to measure owing to goods in the vessel and because in most ships this depth was about half the breadth, D was taken as $B/2$. The result was to 'cheat the rule', and the first example of avoiding a measurement for rating. If the breadth was made smaller on new ships then they actually loaded more cargo for the calculated tonnage. In 1773 there was introduced in England, Builders' Old Measurement, where

$$\text{BOM} = \frac{(L - \frac{3}{5}B) \times B \times \frac{1}{2}B}{85}$$

The $\frac{3}{5}B$ allowed for the rake of the stem. In the United States an almost similar formula was enacted in 1792; known as the 'Custom House Measurement', the only difference was the use of the actual depth instead of the notorious $\frac{1}{2}B$. The Americans thus from the earliest days of their independence were without the effect of a rule which made yachts narrow. For 170 years thereafter (until about 1965), in general American yachts were beamier than comparable British ones!

The first attempts to allow time allowance between yachts of different sizes used the existing tonnage rules (see p. 117), BOM and the Custom House Measurement. Such time allowances were introduced after keen owners had merely built larger and larger yachts to win the cups offered for the following season. In the winter of 1827, Lord Belfast lengthened his yacht *Louisa*, by adding a new mid-section to her. His plan was to beat Joseph Weld's cutter *Lulworth* in racing in the Solent.

The failure of the tonnage rules was that they did not include sail area or displacement in the measurement. That is the reason for the pictures of old-time cutters and schooners carrying great clouds of canvas. Beautiful it might look in the paintings, but it exacted its toll of seamen. Then the hulls were either light with nearly all the ballast in the keel like the narrow *Oona* or light with little ballast like the *Mohawk*. There was also no allowance for freeboard, so the nineteenth-century yachts had very little of this as the many pictures of them show. There were attempts to change the tonnage rules like the Thames Tonnage of 1854 introduced by the Royal Thames Yacht Club, the tonnage equalled

$$\frac{(\text{Length} - \text{beam}) \times \text{beam} \times \frac{1}{2}\,\text{beam}}{94}$$

the length was 'between perpendiculars' – the stem and rudder post – but this hardly helped matters. The last attempt to try a different tonnage rule was in 1882 (by this time the Yacht Racing Association had been formed in Britain by Dixon Kemp and others and was in charge of such things) when the '1730' rule was brought in. In this tonnage was

$$\frac{(L + B)^2 \times B}{1730}$$

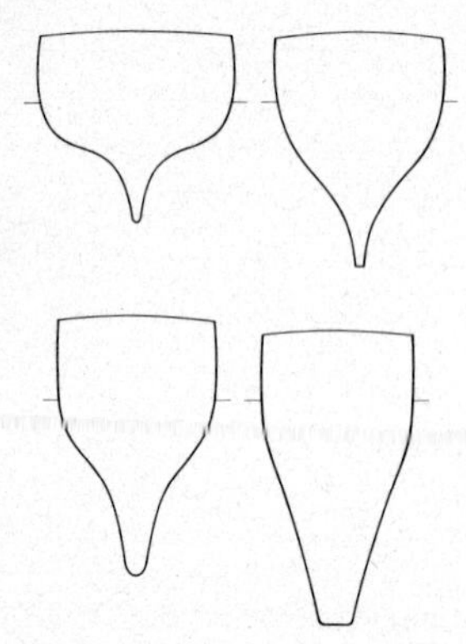

It helped not at all as is shown by the decreasing beam of the yachts year by year. All these yachts were 5-tonners under both tonnage rules. Progressively their LWL and sail area became greater. Extreme draught also became greater, but the beam decreased, as they became more 'plank-on-edge'.

Date	Yacht	LWL ft in	Beam ft in	Draught ft in	Sail area (without topsails)
1873	*Diamond*	25 3	7 2¼	4 6	671
1876	*Vril*	28 4	6 7	5 2	830
1879	*Trident*	32	6	6 3	912
1883	*Olga*	33	5 8¾	6 4	985
1885	*Doris*	33 8	5 7	7	1116
1886	*Oona*	34	5 6	8	1205
For comparison					
1974	*Swan 41*	30 4	11 11	6 6	978

How the mid-section of yachts rating 5 tons changed between 1873 and 1886. All main dimensions increased except that of beam which became the narrowest ever. These boats from left to right are: *Diamond*, 1873; *Trident*, 1879; *Olga*, 1883; and *Oona*, 1886.

The most painful period in yacht rating rules was when the tonnage rules produced ever more freakish craft between 1873 and 1886. In America there were no narrow boats, but the 'Cubic Contents Rule' produced beamy shallow vessels such as *Mohawk*. This rule took the entire volume of the hull below a horizontal plane tangent to the lowest point of the sheer. As in Britain sail area got bigger and freeboards smaller, but beam grew larger and displacement less.

Length and sail area

The greatest controversy over rating rules and yacht design was from 1886 onwards, when changes were introduced on both sides of the Atlantic. The originator of the new ideas was Dixon Kemp. The basis was to tax length and sail area; beam was now left out of the formulae. In Britain there was a 'rating' given by

$$\frac{L \times \text{sail area}}{6000}$$

L was taken at the load waterline and yachts were known as 'raters', such as 2-raters, or 5-raters. The rating was also used to find a time allowance where yachts of different rating raced together. *Britannia*, built in 1893, was measured and came out as a 151-rater. The principal clubs on the east coast of the USA adopted a slightly different 'length and sail area' rule. This was the Seawanhaka Rule, adopted in 1883 and rating was given in feet by

$$\text{Rating (ft)} = \frac{L + \sqrt{\text{SA}}}{2}$$

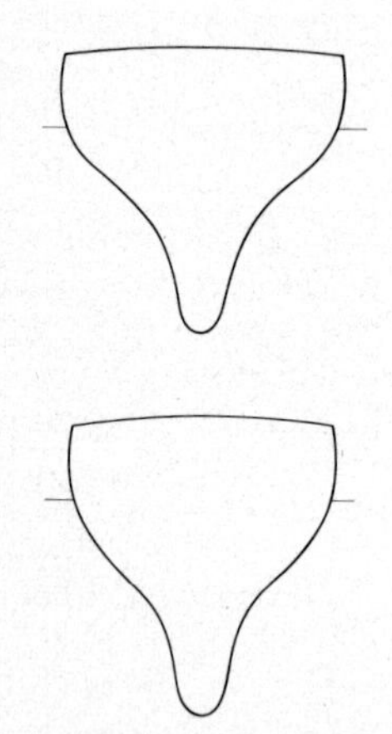

What happened after the notorious tonnage rules were abandoned and a length and sail area rule was introduced: these sections are of *Deerhound*, 1889, and *Varuna*, 1892, which were 40-raters.

The effect of both these rules was to cause designers to try the 'skimming dish'. This had big overhangs at bow and stern because only waterline was measured and U-shaped overhangs gave extra length when heeled and increased speed; it had low freeboard and shallow draught with fin keels. Pictures of the exciting little raters of the 1890s show this. In the large yachts, where there was more expense, extremes were not so frequent (*Britannia* was built to the rule but was essentially a fine yacht), but the most extreme America's Cup boat *Reliance*, designed by Nathanael Herreshoff, was an example of the full exploitation of the length and sail rule. Every year in the smaller classes the boats got longer, which in the formula meant sail area, but these longer boats outclassed the previous existing ones in any breeze. A successful 5-rater increased from being LWL 20 ft in 1883 to LWL 34 ft in 1892.

The 1-rater *Tatters*, a product of the length and sail area rule only waterline was measured hence the long overhangs; there was no penalty for freeboard so it was as low as practicable.

Right
Wee Win, a $\frac{1}{2}$-rater, successful American design in British waters. Still somewhat extreme, her success brought more elaborate rating rules closer at the end of the nineteenth century.

These rating rules by which yachts were measured first began to look more like those of today when it was seen that length and sail area alone were too simple to ensure a sound shape and fair racing. An English 1-rater, *Sorceress* of 1894 owned and designed by Linton Hope with a centreplate had these dimensions: LOA 28 ft, LWL 18 ft 9 in, beam 8 ft 5 in, draught (with plate) 6 ft 5 in, displacement 1470 lb. The most extreme types in England were designed by Herreshoff for British owners – flat-shaped hulls, with bronze centreplates carrying lead bulbs on the end. In 1891 two of this type appeared, *Wenonah*, a $2\frac{1}{2}$-rater, and *Wee Win*, a $\frac{1}{2}$-rater. In 1895 came *Niagara*, a 20-rater. This was too much for the Yacht Racing Association, which adopted in 1896 a new rule devised mainly by R. E. Froude. It introduced 'G' which was a 'skin girth' – a measurement close to the midship sections. The more pronounced the fin, the bigger would G be. In this the rating was expressed in feet

$$\text{Rating (feet)} = \frac{L + B + 0{\cdot}75\text{G} + 0{\cdot}5\sqrt{S}^{*}}{2}$$

This failed to stop the skimming dishes, so in 1901, a new factor 'd' was brought in which was the difference between the 'skin girth' and the 'chain girth' and much more reflected the light displacement section. The 1901 rule was

$$\text{Rating (feet)} = \frac{L + B + 0{\cdot}75G + 4d + 0{\cdot}5\sqrt{S}}{2{\cdot}1}$$

The skimming dish problem in America was solved by Nathanael Herreshoff, poacher turned gamekeeper, who devised the Universal Rule of 1901 for the New York Yacht Club. He brought in displacement (in cubic feet) and measured L, one quarter of maximum beam from the centreline, to reduce scow-like ends. If sail area (square rooted) exceeded length by more than 35 per cent, there was a heavy penalty. The Universal Rule was

$$\text{Rating (feet)} = 0{\cdot}2\,\frac{L \times \sqrt{S}}{\sqrt[3]{\text{Displacement}}}$$

The Universal Rule with amendments and extra 'small print' was used with success for inshore racing in the United States until 1924. After that time it was used only for the biggest inshore racing yachts, in effect the J class of America's Cup contests, rating under this rule, at 76 ft.

* S = Sail area. As it is a square measure (e.g. square feet), it is usually in these formulae as a square root.

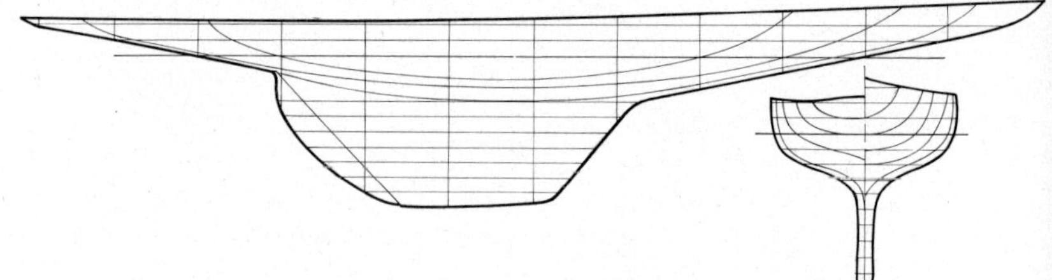

Reliance, defender of the America's Cup in 1903, was designed to the Universal Rule. Though displacement counted, the rule still produced very long overhangs, so that the forward one was 27 ft 8 in. Over-all length of the yacht was 143 ft.

The phenomenon of numerous clubs in different countries struggling to construct a rating rule that would not be broken by the designers was confined to the years just before and after the turn of the century. Since then there have been principal national and international rules though several organizations have had their own systems of measurement. The great fallacy of this period was that a simple formula could be devised to solve the problem of equitable yacht rating. It only dawned slowly on the Victorians that there must be a number of parameters introduced (length, beam, sail area, freeboard, depth, etc.) and these must also have limits and qualifications. In the yachting press of the time, yachtsmen who failed to grasp this, debated endlessly as to whether there should be 'a sail area rule' or 'a displacement rule' or some other major dimension – but all such dimensions had to be considered. One of the national rules was that of the 'Union of French Yachts' (since superseded by the French national authority). Its significance today is that it is the original rule under which the One Ton Cup (see p. 43) was sailed. The little boats of LWL 17 ft of 1 ton rating had to rate at 1 under this formula, first introduced on 5 November 1892 and amended in 1899.

$$\text{Rating} = \left(\frac{L-P}{4}\right) \times \frac{P \times \sqrt{S}}{1000\sqrt{M}}$$

P was the length of the perimeter comprising the chain girth and the deck. M was the area of the midship section in square metres, measured with horizontal ordinates equally spaced on the half beam.

From left to right
6-metre class yachts in England racing in 1911, soon after the introduction of the first international rating rule.

What 6-metre yachts looked like in 1923. Bermuda rig was adopted but the mainsails were large and genoas had not been invented.

First international rule

The first international rule was decided at a meeting in London in 1906 initiated by Brooke Heckstall-Smith, Secretary of the Yacht Racing Association. It is the most important year in the development of yachting as an international sport. Sixteen countries (but not the USA, although invited) resolved to adopt the International Rule (its correct title), which can be seen to derive from the YRA Rule of 1901. At last freeboard (F) was introduced (though it was still low by today's standards). In deference to the European countries the rating was defined in metres.

$$\text{Rating (metres)} = \frac{L + B + \frac{1}{2}G + 3d + \frac{1}{3}\sqrt{S} - F}{2}$$

Although time allowances were possible, the plan was to have classes racing level at 5, 6, 7, 8, 9, 10, 12, 15, and 23 metres. Boats were built to all of these, but the most popular were the 6-metres, 8-metres, and 12-metres. The famous metre boats were born. One of the first major trophies to be transferred to boats of the new formula was the One Ton Cup for yachts rating at 6-metres. Heckstall-Smith gave the following as equivalents in size under the various rating rules spanning 50 years in England. For a yacht of about 50 ft on the waterline:

1881–6	20-tonner	
1886–95	20-rater	(the length and sail area rule purposely made the value about equal)
1896–1906	52-footer	(the Froude rule brought the rating back to approximate LWL)
1906–	15-metre	(a bit less than the LWL and given in metres)

The longest lasting rule of rating has been the International Rule of the International Yacht Racing Union first adopted in 1906. True the rule has been extensively modified, but boats have been built to it and its modifications continuously for nearly 70 years. It is also true that it is only actively used for 12-metres, but in the important matter of America's Cup. A few

By 1937 6-metre yachts had genoas, shorter booms and simplified masts.

A modern 6-metre sailing at Seattle.

6-metres are built to it (see p. 69), the historic irony being that these are in the north-west United States: yet the USA at first refused to have anything to do with the metre boats of England and Europe. In England they last raced in 1955: a few remain on the Swiss Lakes. But it is a handful – one 6-metre to every 900 International Offshore Rule class yachts.

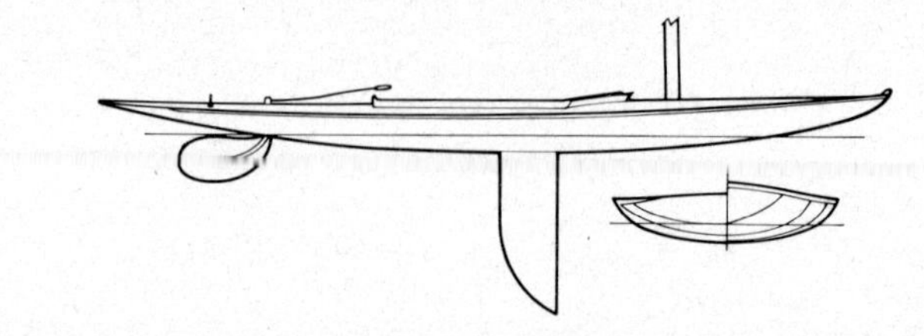

vned and .inton Hope as 394. She was skimming y the length ule. Only sail area were m grew, d displacement so below, sailing.)

The second great era of the metre boats began after the First World War, when the USA joined in talks about international classes. The results were the adoption of the metre classes in the USA for smaller vessels and the Universal Rule for big yachts – in effect the J class. The International Rule became the widest used rule to date. Now the Seawanhaka Cup became used by 6-metres and the British-American Cup. The strength of the metre boats were regulated by scantlings drawn up by *Lloyd's Register*. After the First World War the rule was modified to the following

$$\text{Rating (metres)} = \frac{L + \frac{1}{4}G + 2d + \sqrt{S} - F}{2{\cdot}5}$$

Beam was dropped because under the old formulae less beam meant a lower rating and yachts were again getting narrower – no, they had not learnt! A minimum beam rule was brought in instead. In 1933 the International Rule was changed to the form in which it is used today for the 12s.

$$\text{Rating (metres)} = \frac{L + 2d + \sqrt{S} - F}{2{\cdot}37}$$

The term G had disappeared as the girths were now incorporated into the methods for measuring length. There are stipulations on minimum beam, limitations on such dimensions as freeboard and draught and the height of the head of the headsail and spinnaker up the mast (it cannot go to the masthead). Total height of the rig above the deck is limited to twice the rating figure plus 1 metre and the foretriangle height is limited to 75 per cent of that. Time allowances were once (before 1940) issued for use with the International Rule, but never generally adopted.

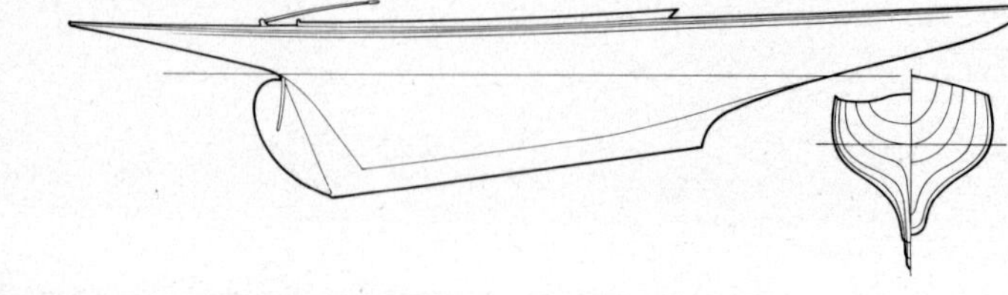

The profile and body sections of a 6-metre of 1912 designed by Morgan Giles. A high-peaked gaff rig was used on this hull which was 35 ft and LWL 20 ft 6 in.

International Offshore Rule

The only rating rule of any significance in the world of yacht racing today is the International Offshore Rule. This is the final unification of all the rules of earlier years: there will no doubt be new versions of it, but local and national rating rules are behind us. A further new rule would be an international one. The IOR is for habitable single-hulled yachts (in 1975 there was just available an international offshore multihull rule but it was at

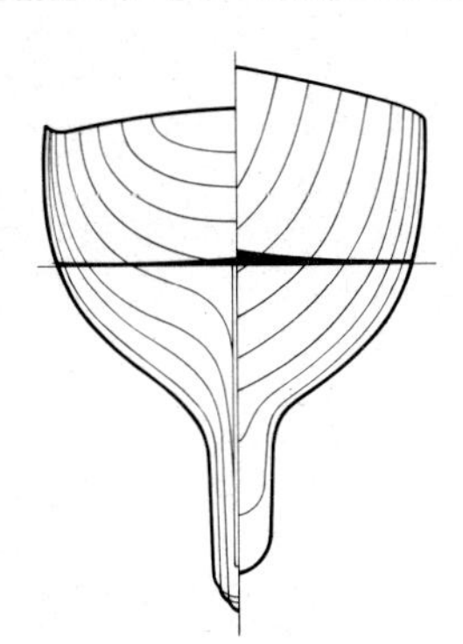

6-metre sections of 1936, designed by Bjarne Aas. Compared with the 1912 boat, the class had become narrower and longer (LOA 37 ft 3 in and LWL 23 ft 7 in). This was because B was back on the top line of the formula: a minimum beam rule had been introduced to stop the re-emergence of plank-on-edge boats.

an early stage of development) and not for the day-racing boats of the first half of the twentieth century and before. The IOR is a complex one in that a large amount of supporting restrictions and penalties are necessary for each factor in the main formula. Each factor is itself derived from many sub-measurements. Length for instance is a matter of measuring the shape of the ends. Draught has provisos for centreboards and types of rudder; every sail has a way of being measured and various types and parts of sails are forbidden and penalized.

The evolution of the IOR is important because it contains elements of the lessons learned in all the rules for a hundred years. The first drafting was in England in 1912 when a number of yachtsmen formed a 'Boat Racing Association' to race smaller yachts than were then encouraged by the Yacht Racing Association. As was the custom then, it deemed it right to have its own rating rule, the BRA rule. This proved successful in rating a variety of existing boats, and although an 18 ft class rating to the rule was proposed the First World War intervened. The BRA rule was

$$\text{Rating (feet)} = \frac{1}{3}\frac{L \times \sqrt{S}}{\sqrt[3]{\text{Dis}}} + 0{\cdot}25L + 0{\cdot}25\sqrt{S}$$

It will be seen that the first half of the formula is essentially the Herreshoff Universal Rule and the second half is the old Seawanhaka Rule. When the organizers of the first Fastnet race in 1925 wanted some sort of rating rule for the small fleet of existing cruisers that took part, they sought advice from Malden Heckstall-Smith (brother of Brooke) who recommended something close to the 1912 BRA formula. Meanwhile, in the USA, the early Bermuda races were conducted merely by giving a time allowance on length. Again, such a procedure worked if the boats were roughly similar and not sailed highly competitively – but it did not work well enough, because for the Bermuda races from 1928 to 1932 and for the 1926 Fastnet race a slight variation of the Universal Rule was used. It did away with displacement which was very difficult to measure and used a combination of beam (B), and a concept from the tonnage rules of a century before, the depth (D) of the hull (not the draught, but the 'cargo space'; for example from the deck to the bottom of the usable space in the boat. But note the definition as used today has become highly refined, as have progressively all other factors used). So the early Fastnet/Bermuda race formula was

$$\text{Rating (feet)} = \frac{0{\cdot}2L \times \sqrt{S}}{\sqrt{BD}}$$

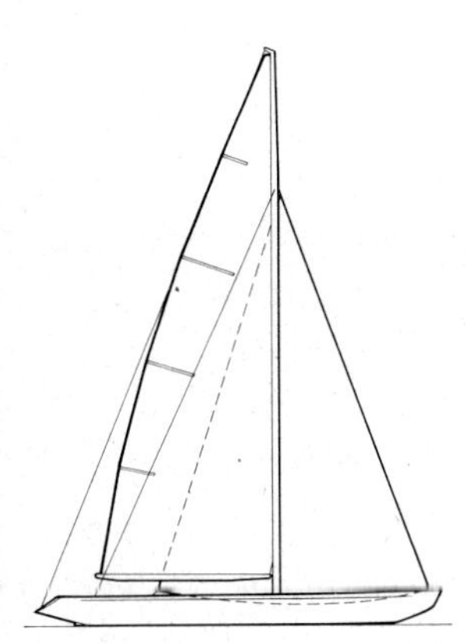

A 6-metre of 1975 designed by Douglas Peterson (US). Note low boom and snubbed ends. The only regular racing for modern 6-metres is at Seattle.

Transatlantic agreement on this formula did not last long. The Bermuda race organizers, the Cruising Club of America, from 1934 adopted a different rule, which had a strong likeness to the Seawanhaka Rule. But as well as having length and sail area, there were a whole series of small corrections which were found by seeing the differences against a 'base boat' – an ideal decided by the rule-makers – so the first CCA rule looked like this

$$\text{Rating (feet)} = 0{\cdot}6\sqrt{S} \times \text{rig allowance} + 0{\cdot}4(L \pm B \pm D \pm P \pm F + A + C)$$

Apart from many minor differences, the main technical cleavage between this American and the British rule for measuring ocean racing yachts, was the system of determining length. The British were using girth measurements, that is finding the point to take on the bow and the stern by putting a tape round them relating to their bulk. The Americans considered a profile only and took the length 4 per cent of the waterline length above the

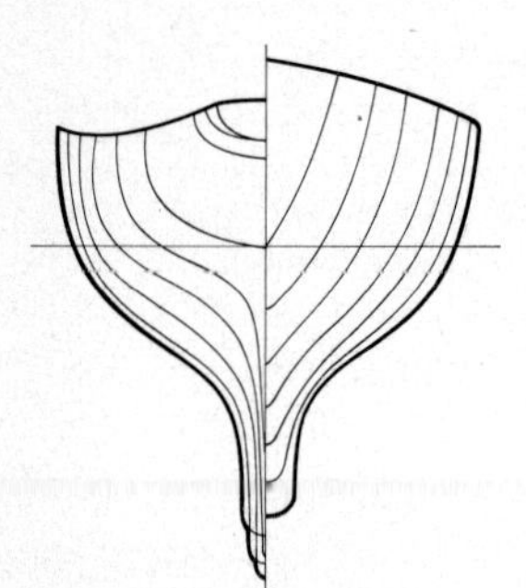

The sections of a 1932 ocean racer to the RORC rule designed by Olin Stephens. The lines were used for the British boat *Trenchemer*.

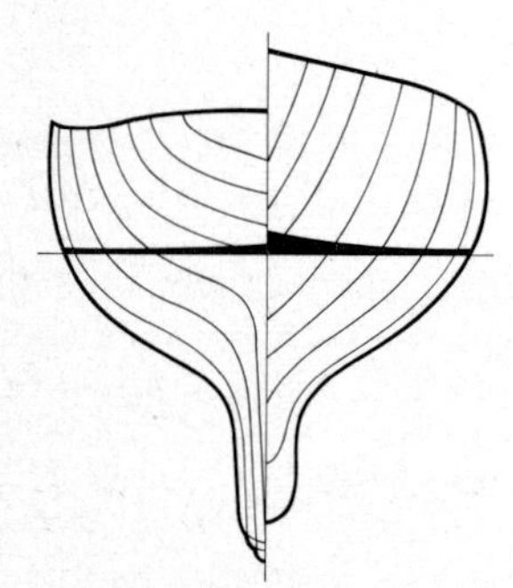

Hull sections of a 1949 ocean racer to the RORC rule. This typical 50 ft 6 in hull, *Northele*, still aped metre-boat hulls.

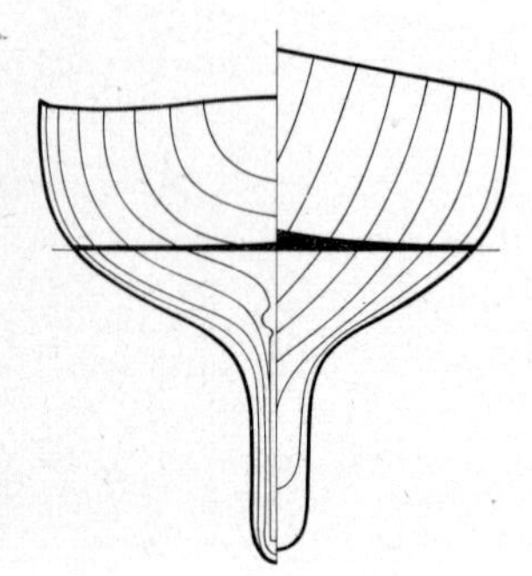

Hull sections of a 1965 ocean racer to the RORC rule, before it was replaced by the IOR. Hulls were now lighter and much beamier than earlier boats to the same rule.

water plane; the result was their sterns could be wider than the British boats who had to pay for such a shape that gave more concealed sailing length.

The CCA used to adjust the values of the differences on the 'base boat' very frequently, as the Bermuda race results showed an advantage to any one type. In its final form, the CCA rule looked like this

$$\text{Rating (feet)} = (L \pm B \pm D \pm \text{Displacement} \pm \sqrt{S}) \times \text{Stab } F \times \text{Prop}$$

Displacement had been brought in by weighing the boats, Stab *F* included the stability in the sense of resistance to heeling under sail, and Prop indicated a series of allowances for different types of propeller installation when sailing. In 1970 the CCA abandoned its rule in favour of using the IOR, which included many features of the CCA rule in its make-up.

In Britain, the Royal Ocean Racing Club settled on its own rule in 1931, which kept its basic form until phased out in favour of the IOR on 30 September 1970. Of course, there were numerous revisions in the way that the principal factors in the main formula were arrived at: the last major revision of the RORC Rule was in 1967 and the rule looked like this:

$$\text{Rating (feet)} = 0{\cdot}15\,\frac{L \times \sqrt{S}}{B \times D} + 0{\cdot}2(L + \sqrt{S}) \pm \text{Stability allowance} - \text{Prop} + \text{draught penalty}$$

The stability allowance depended on the weight of the engine and the material of the keel and mast, the propeller allowances were the same as the CCA, and the draught penalty controlled excessive draught. The CCA Rule was the principal rule in North America but there were others; the RORC Rule although entirely controlled from London by the club from which it got its name was by the 1960s used for all important events in Europe, Australasia, and the Mediterranean.

5·5-metre rule

The only rating rule to be introduced after 1945 and to be widely used internationally was the rule of the 5·5-metre class. This was a new day-racing keel-boat class introduced by the International Yacht Racing Union in 1950. The IYRU declared it an Olympic class instead of the 6-metre and it was slightly smaller than the latter. However, the use of a formula in this way for day racing was seen in the 1950s to belong to a past era and the class was removed from the Olympic list in favour of the one-design Soling keel boat. The 5·5-metre is now obsolescent. The formula used is of interest as it was proposed by Malden Heckstall-Smith and Charles E. Nicholson. One of the features of the 5·5-metres was that no Genoa jib was allowed for in the rules, the device being to measure the actual area of the jib and not allow 'free' overlaps like the ocean racing rules and the old metre-boat rules. The rule was, like the RORC Rule, a combination of the Universal and Seawanhaka formulae

$$\text{Rating (5·5 metres)} = 0{\cdot}9\left\{\frac{L \times \sqrt{S}}{12 \times \sqrt[3]{\text{Displacement}}} + 0{\cdot}25L + 0{\cdot}25\sqrt{S}\right.$$

The boats were of a narrow type because the intention was for a single rating only and for the class to race on exactly equal terms. There were minimum and maximum dimensions to keep the intention for beam, draught, displacement, sail area, and so on. Even sheer, tumblehome, and the number of crew (3) were controlled. Between 1945 and 1970 there were other rating rules in use in the world, but these were of a minor nature.

Dyna, a typical CCA rated yacht of the 1960s.

Right
Jocasta, a typical RORC rated boat of the early 1950s.

Below
The 5·5-metre class, introduced by the IYRU in 1950. It was the last international inshore formula class.

There were local rating rules on the Baltic and, as mentioned, in various parts of the USA. They included the KR Rule, the Storm Trysail Club Rule, and the Offsoundings Club Rule. One rule actually introduced by the IYRU in 1949 attempted to create a number of cruiser racer classes at 7, 8, 9, 10·5, 12, 13·5, and 15 metres to supersede the old metre boats. But this was doomed to failure because hundreds of ocean racers, which is what the cruiser racers would have been, were already racing to the CCA and RORC rules. There were some of these boats built and a few sailed in the Scandinavian countries and an 8-metre cruiser-racer class sailed on the Clyde until 1970.

IOR

The death of all such classes was sealed by the adoption of the International Offshore Rule by the RORC and CCA in 1970: when these two clubs which have hundreds of yachts measured to their rules gave up their own autonomy in the matter, yachtsmen in all countries followed suit. This remains the position today: in 1974 there were 10 000 yachts measured to the IOR and holding valid certificates. This is far in excess of any keeled, or cruising, or ocean racing class ever. The IOR has evolved as a rule from the lessons of about a century – ever since the old tonnage rules were abandoned in Europe and America. It is, therefore, the culmination of the rating and measurement of yachts. No rule has ever been controlled so tightly on an international plane, small faults receiving early attention from a small technical committee, which reports to the controlling body, the Offshore Rating Council, which meets every year. No rule has ever been used for so much variety in racing. It is for self-righting, single-hulled, habitable sailing-yachts between 16 ft and 70 ft of rating (say LOA 23 ft to LOA 85 ft) and using a number of different time-allowance systems (Section F), yet is used for the following types of race:

(*a*) Regatta inshore courses
(*b*) Ocean races including overnight races
(*c*) Trans-ocean races
(*d*) With time allowances between yachts of widely differing sizes
(*e*) Without time allowances between boats of certain ratings: 18 ft, 21·7 ft, 24·5 ft, 27·5 ft, and 32 ft – these are the 'Ton classes'.

The shape of a boat designed to the International Offshore Rule in 1975. Successful ocean racer *Bootlicker* has ample accommodation but is also designed to get the best speed for her rating. Note a clean deck without the various equipment of the older yachts.

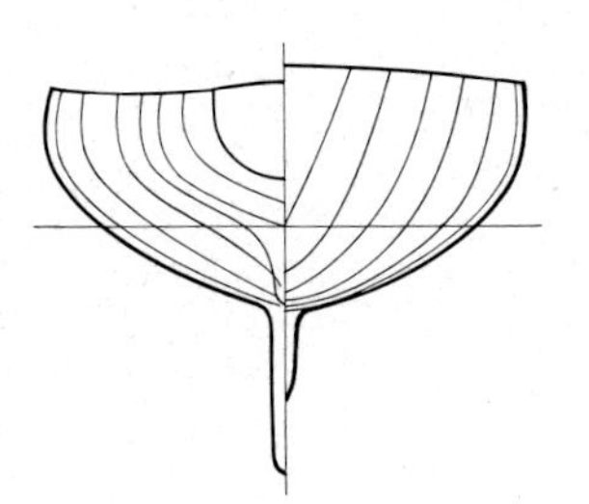

The sections of a 1973 ocean racer designed to the IOR: beam and pronounced fin are characteristic.

The IOR is unprecedented in that now almost every day someone in the world is using it for racing. It is the most complex rule, filling a 60-page booklet. The first CCA rule covered $3\frac{1}{2}$ pages; the first length and sail area were a few lines, but now there are scores of sub-measurements before the main factors are determined on a computer print-out. The rating certificate since 1970 has been in the form of a computer print-out. The basic formula, however, has not changed a great deal. Its descent from the RORC Rule is evident, its CCA ancestry is more in the systems of measurement used, especially in regard to sail area. The old Universal and Seawanhaka rules are still very much there. The rating is calculated in both feet and metres; feet are primarily used because the rule is based on the British and American rules. The rating of the yacht is given in feet and tenths which means there are exactly 541 different ratings between the laid-down 16·0 to 70·0 feet rating limits. The IOR formula is:

$$\text{Rating (feet or metres)} = \left\{ 0{\cdot}13\frac{L \times \sqrt{S}}{\sqrt{(B \times D)}} + 0{\cdot}25L + 0{\cdot}2\sqrt{S} + DC + FC \right\} \times EPF \times CGF \times MAF$$

In this formula D is depth, DC is draft correction, FC is freeboard correction, EPF is engine and propeller factor, CGF is centre of gravity factor, and MAF is movable appendage factor.

A reminder of how the rating rule is used today

When each yacht has been measured to the International Offshore Rule and the results entered in the computer and a rating in feet and tenths of a foot obtained – what happens then and how does this aid in enjoyable racing of the yacht?

As just indicated, the certificate of rating shows that all the factors of the rule have been measured on the yacht and as a result she has a potential speed which is a theoretical length. If any alteration is made to anything measured, then the certificate becomes invalid. This might be caused by alterations, even small ones, in the size of sails, height of the mast, type of propeller, moving the ballast of the keel, or altering the trim of the yacht by some other means. But if the essentials are kept inviolate then the yacht has a rating of say 25·2 ft. The metric equivalent would be shown officially on the certificate. Continental boats coming to race in Britain or America might have a metric certificate but the rating would be there in feet for the race committee to use. Incidentally French and Italian clubs like to use feet for rating nomenclature, because it differentiates from the actual length of the yacht which for them would be in metres. Feet for rating are rather like 'hands' in which horses are measured and even when metric measurement becomes generally used in English-speaking countries, the rating is likely to remain in feet.

With the certificate in his pocket, the owner can enter the yacht for any race in which it is announced that the event will be under the IOR and a suitable time-scale or at a level rating. All the tedium of the measurement and the computation is over and all the club is interested in doing is getting the rating figure against the name of the owner and his yacht on the race-card. The rating also enables the organizing club to put the yacht in the appropriate class, grouped by rating. There might be one from 19·5 to 22·4 ft and another class from 22·5 to 26·4 ft. The yachts start by classes or altogether. If the race was on level rating, for instance, for Half Ton Cup boats, then the first home wins. It is merely up to the owner to make sure that the yacht has taken advantage of the rule to get the rating up to 21·7, the Half Ton limit.

Bicolour navigation light.

More likely the fleet is a mixed one. Then the time of each finisher is taken. The elapsed time of each finisher is quickly reckoned by subtracting the time of the start and a correction is then applied to it. This correction comes directly from the rating, and is the time allowance (see Section F) in the form of seconds per mile or seconds per time sailed. The result is corrected time; the yacht with the shortest corrected time is the winner; the yacht with the second-best corrected time is the second, and so on. Yacht A with a higher rating than Yacht B, must finish ahead of her to have a chance of beating her. Not only that but A must finish far enough ahead of B to 'save her time'. If B finishes at the same time as A, B has clearly beaten her. Or if B can finish very soon after A, she may well have beaten her. But the use of an international rating means that the figure, found by the most refined method ever used in yacht measurement is available for enabling the yacht to race against any other yacht similarly rated.

Breakthroughs

The number of yacht designs that have been actual breakthroughs, or landmarks in the history of yacht architecture, is very limited. Those of the last century have already been mentioned in this section – *Jullanar*, *Gloriana*, *Britannia*. Boats like *Wee Win* have been more instrumental in having rating rules changed rather than advancing design. But sometimes a design which has found a loophole in the rule has also pointed a way to a new concept in design. Other cases have advanced design by one feature or another, without torpedoing the rating rule. Yet when the rating rule is a bad one, the good idea may be prevented from finding expression – as in the case above of *Evolution*. A final category is that of the freak. This might be defined as an attempt to break through to racing advantage by highly unusual features intended to increase speed or take advantage of the current rating rule. Such a boat is often undesirable in terms of safety at sea, or even by current feeling as to how a sailing boat should look. By a freak is meant a boat about which there is plenty of talk, but which fails to impress when racing: its concept of design is not then pursued further.

The design progress in modern racing boats is most apparent the nearer to the craft one goes. Because 'a boat is a boat' some features at a distance may appear to have much in common with those yachts of 50 or even 100 years ago. Indeed such old boats are still in commission and, when they sail past, do not attract as much attention as would a vintage car of only 50 years age, when seen on the modern road. (The progress in cars with a mass market and vast investment has obviously been much quicker.) The contrast in materials and fittings between first-class yachts designed 25 years apart shows the major factors, outside the rating rules, which have contributed to changes in design.

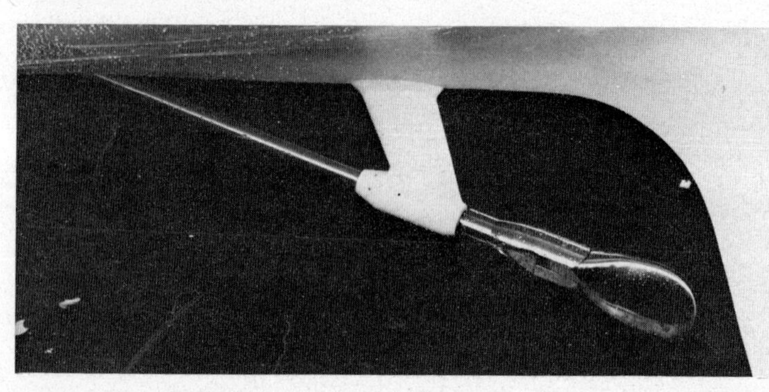

Folding propeller.

Feature or material	Yacht of 1946	Yacht of 1971	Significance of modern usage
Hull material	Wood planked	GRP Light alloy High tensile steel Foam sandwich Cold moulded wood or planked (rare)	Rigidity, required weight to give light displacement, difficult shapes can be moulded GRP does not rot
Sails	Egyptian cotton	Dacron (Terylene) or nylon	Rot-proof, controlled shape, strength, handling easier, lighter weight for heavy weather sails, long life
Spars	Wood, hollow or solid	Light alloy anodized	Negligible maintenance, lighter weight, no rot
Deck Fittings	Bronze, Gunmetal, with Lignum vitae, Galvanized steel	Stainless steel, light alloy, laminated plastic, solid nylon and tough plastics, plastic coatings	Numerous opportunity for design, light weight, easy lubrication
Rope	Hemp, Manilla (natural fibres)	Terylene, nylon Polypropylene (synthetic fibres)	Much longer life and strength good appearance
Cabin fabrics	Cotton, etc. on hair	Waterproof synthetics foam interiors	Comfort and drier below
Food storage	'Deck safe' louvred locker (Europe)	Ice-box (used in USA for far longer)	Cleaner, more space available
Propellers	Solid and feathering	Folding and projected from inboard/outboard unit available	Better sailing performance for rating, but power results vary
Navigation lights	Probably paraffin in shrouds	Sealed electric on pulpit and masthead	More likely seen and dependable
Application of manual power for sheets and halyards	Blocks and tackles, some small bronze or custom-made winches, wire used on them on big yachts	Range of needle bearing or oil-bath multi-geared winches of all sizes synthetic cordage used on them	Enables big sails to be sheeted, much increased loads, fast tacking and handling of spinnakers, safer with wire or large thrashing blocks
Safety equipment	Kapok life-jackets, wood dinghy, flares	Inflatable life-rafts, harnesses, lifelines, RT	Vast increase in safety conciousness

Feature	1946 yacht	1971 yacht	Remarks
Paints and coatings	Mainly intended for wood. Good range specially for yachts including anti-foulings and enamels	For all surfaces from specialist firms	Still change from time to time. Early hopes of polyurethanes not realized. Anti-foulings still need renewing and preparation still more important than any chemical
Standing rigging	Wire (soft fibres disappeared in previous century) Ends spliced	Variety of including solid rod and special terminals	Principles the same but more loads induced. Big improvement in subsidiary fittings
Ballast keels	Lead or iron bolted on bottom. Some boats with pigs of inside ballast	Same, but sometimes encapsulated in GRP and alloy hulls. Some makers can cast lead very densely	Nothing with better specific gravity than lead. Uranium banned for racing

The list could go on giving many details, but the above are important and typical. Here, however, are a few which have changed perhaps in brand names and techniques, but not very greatly in historic terms

Influential designs

The first design to show British yachtsmen that racing on the open sea need not be in heavy cruising and workboat types was *Nina*, owned by Paul Hammond. She was designed by Starling Burgess for a transatlantic race to Spain in 1928 and won the Fastnet race the same year, beating all the British entrants. She was the first American yacht to do so, but by no means the last. The design was only radical in that it carried the tradition of the fast Baltimore schooners, with fine bows and lines generally and unnecessary weight omitted, into yachting in England. The schooner rig had the mast rather far forward because her sail area had been measured in the USA under the Universal Rule. Down below, *Nina* was generally open, unlike the much-bulkheaded cabins of the other yachts. Dimensions of *Nina* were LOA 59 ft, LWL 50 ft, beam 15 ft 2 in, draught 9 ft 3 in, sail area 2300 ft^2. She was the first yacht to sail or race across the Atlantic with a Bermuda rigged sail. Whatever the variation in rules, *Nina*, under new ownership, remained competitive. In 1940 she took part in the last offshore race to be held on salt water before the Second World War put a stop to it – this was from Block Island round Mount Desert Rock, finishing at Gloucester, Mass., 145 miles. *Nina* is the oldest yacht ever to have won the Bermuda race, having done this in 1962 under the ownership of De Coursey Fales, who was then 74.

The yacht with the greatest influence on ocean-racing design before 1940 was *Dorade*. Like *Nina*, she was American designed and owned – by Rod and Olin Stephens in 1930. Like Starling Burgess, Olin had America's Cup success ahead of him. Since Rod and Olin's firm, Sparkman and Stephens of Madison Avenue New York, is still turning out designs of remarkable

In the mid 1970s the 'blooper' or 'big boy' has developed additionally to the spinnaker to give extra sail area off the wind

One of the last yachts built to the RORC rule before it was absorbed into the IOR, *Roundabout*, owned by Sir Max Aitken

yachts of every type four decades later, they are rightly considered the world's leading designers of racing yachts. No other firm of designers approaches them for having as many yachts in successive years to win prizes in international events.

Although *Nina* was designed and equipped for ocean racing – and in particular for the Transatlantic race – she followed the theme of a successful traditional American type. *Dorade*, however, was to be an all-round ocean-racing yacht to the current rules (then the same in the USA and Britain). After her great successes (Fastnet race 1930 and 1931 and Transatlantic race 1931 – all by large margins), British yachtsmen at last began to build racing yachts for offshore use instead of trying to sail heavy old cruisers. American yachtsmen flocked to the Stephens brothers for new designs after they had returned from England to a ticker-tape welcome in New York, the only yacht designers to have received such an honour.

Dorade was 52 ft, LWL 37 ft 3 in, beam 10 ft 3 in, draught 8 ft, sail area 1100 ft^2. She was a Bermuda yawl and put an end to the use of gaff rig for ocean racing; after her the schooner rig, so common in America, also began to die out for serious racing (in 1974 among 10 000 IOR class yachts there were 3 or 4 schooner rigged). An interesting aspect of design evolution was that *Dorade* owed something to the metre boats of the International Rule. Olin Stephens sailed some of these boats on passage between Halifax, N.S., and Long Island Sound (they were fairly new to America then) before *Dorade* was designed and was impressed by their behaviour offshore.

After *Dorade*, the new designs from British designers for their clients had the *Dorade* look. Designers included William Fife, John Tew, Robert Clark, Laurent Giles, and Charles E. Nicholson. Nicholson's *Bloodhound*, a yawl of 63 ft 5 in, built for Isaac Bell in 1936, was very like the American boat in line and rig. The inherent usefulness of the type is demonstrated by the fact that she was bought by the Queen and Prince Philip in 1961 and, after having the rig and interior modernized by John Illingworth and Camper and Nicholsons, was used for racing and cruising for ten years, before being sold again. Since then the yacht has cruised and chartered.

Not all British sailors acclaimed the ability of *Dorade*: an article in *Yachting World* in 1933 said: 'The modern ocean racing machine represented by such as *Dorade*, has not been produced simply by retaining and developing the virtues of the cruiser and blending thereto something of the speed of the racing yacht. The type has vices of its own. It requires a large and well-trained crew to sail these yachts, sacrificing as they do, comfort and even safety for the sake of speed. Their deep outside keels give them weatherliness and speed at the expense of seakindliness. . . .'

Rod and Olin Stephens built *Stormy Weather* as a successor resembling *Dorade* and returned in *Stormy* in 1935 to win the Fastnet yet again.

The first 12-metre before the Second World War of this largest class to the International Rule was *Vim* designed by Sparkman and Stephens. Her significance is that she was the link between the regular 12-metre racing at the end of the 1930s and the revival of 12s for America's Cup only in 1958. She was built in 1938 for Harold S. Vanderbilt, as his large racing yacht, after the J class had raced for the last time. Her dimensions were LOA 70 ft, LWL 45 ft 3 in, beam 11 ft 9 in, draught 9 ft 1 in, sail area 1817 ft^2, displacement 60 000 lb. In 1939 she spent a season in England racing against the British 12s like *Tomahawk* (T. O. M. Sopwith) and *Evaine*. There were then 12s in France, Germany, Italy, and Sweden, as well as Britain and the USA, but the J class owners were now in 12s.

In 1958 *Vim* was put back into racing trim, having been temporarily a cruiser, by a New York Yacht Club syndicate. In the America's Cup trials

Nina.

Right
Tre-Sang, 30 square metre class which Hasler used for ocean racing in 1946.

she began consistently beating the new boats, though finally Olin Stephens's development of her, *Columbia*, got the upper hand. Thus her proved ability forced the new American 12s into shape and ensured the modern run of NYYC success in the Cup. The subsequently defeated *Sceptre* had in trials in the same year difficulty in beating *Evaine*.

When the Royal Sydney Yacht Squadron decided to challenge for the Cup, *Vim* was chartered and brought to Sydney. She was, therefore, a departure-point for the Australians where no 12-metre (or even an 8-metre) had ever been raced before. The formidable *Gretel* was the result.

The most impressive demonstration that very light-weight yachts were really suitable for use at sea was by *Tre-Sang* in 1946. In the year when regular ocean racing was restarting Colonel H. G. 'Blondie' Hasler and a crew of two won the 'small class' championship through consistent sailing in the Channel and Irish Sea in all weathers. *Tre-Sang* was never built as an ocean racer. She was one of the 30 square metre 'skerry' class, raced inshore between the wars and originating in Sweden. These were low freeboard boats, with immense overhangs and light displacement, but Hasler despite really roughing it in the tiny cabin showed that massiveness was not needed at sea. The dimensions of *Tre-Sang*, designed by Knud Reimers of Sweden, show the strange proportions: LOA 42 ft, LWL 27 ft, beam 7 ft 2 in, draught 4 ft 9 in, displacement 6225 lb, sail area 323 ft^2 (30 metres^2).

The sailing-yacht to have the greatest effect on design since the last war in British waters and to a great extent beyond was *Myth of Malham*. She was designed by Laurent Giles and Partners to the ideas of Captain John Illingworth, RN, who owned the yacht and in her scored an amazing series of successes from 1947, her first season when he won the Fastnet race. The dimensions were LOA 37 ft 9 in, LWL 33 ft 6 in, beam 9 ft 4 in, draught 7 ft. Like *Tre-Sang* the displacement was light, but the ends were short. Long ends served no purpose at sea or under the rule. After *Myth of Malham*, there were boats extremely like her, while almost all others had one or other of her features. In terms of her effect on design these were:

Myth of Malham.

High freeboard: On *Myth of Malham* this was partly a rule-cheating device because the the RORC Rule in 1947 measured 'depth' (*D*), in the formula, from deck to hull. The rule was changed subsequently, but it was seen the high freeboard was not a disadvantage.
Mast well inboard: Masts have since tended to be closer to the centre – fore and aft – rather than forward, giving large headsails and small mainsails. The rig in future was all inside the ends of a boat.
Light displacement: Boats have been built with heavy displacement (a comparative term) since, but always the attempt has been to keep the hull as light as possible in accordance with strength and put extra weight into the ballast keel to give stiffness and power to carry sail.

Two features of yachts today which were not present in *Myth of Malham* were the separate rudder well aft of the fin keel and more beam. The first was tried by various designers including Illingworth in *Wista* and *Mouse of Malham* in 1954–5, but did not gain general acceptance until later. The second was eschewed by British designers for no other reason, that the author can think of, than prejudice. Certainly the RORC Rule did not discourage beam. C. E. Nicholson who lived until 1962 was always quoted as saying 'Nothing stops a boat like too much beam.' English boats, it was said, had to be narrow to beat against the Channel chop. It was not true. Today's proportions of beam (a boat of *Myth*'s LWL would have 12 ft 6 in beam) were not evolved in England until about 1964. *Myth of Malham* foundered and sank without loss of life in French ownership off the French coast in 1972.

The first demonstration that really small yachts could race offshore was by *Sopranino*. She was a racer in miniature and in 1950 sailed with RORC fleet from England to Spain, then raced in the English Channel with the newly formed Junior Offshore Group for yachts below RORC size (break-point LWL 24 ft), and finally crossed the Atlantic by the southern route, manned by Patrick Ellam and Colin Mudie. John Illingworth had a hand in her concept, and the designers were Laurent Giles and Partners, she was wood clinker built and her dimensions were LOA 19 ft 8 in, LWL 17 ft 6 in, beam 5 ft 3 in, draught 3 ft 8 in. Other smaller boats had cruised extensively and crossed oceans, but the design of *Sopranino* showed that the growing modern techniques of fast sailing in bad weather could be applied to really small yachts (look at the size of the other yachts mentioned above).

The design to have the greatest effect on yachts to the Cruising Club of America Rule was *Finisterre*, owned by Carleton Mitchell. She was first overall in the Bermuda race in 1956, 1958, and 1960, a record which no other yacht in this race had approached. She also scored many other successes. Her dimensions were LOA 38 ft 8 in, LWL 27 ft 6 in, beam 11 ft 3 in, draught 3 ft 11 in (7 ft 7 in with place down), displacement 21 500 lb, sail area 713 ft². So she was not only a heavy centreboarder, but had great beam for her time, and was a yawl. The keel was long and straight. Construction was planked mahogany. This Sparkman and Stephens 1955 design helped to ingrain beam and two-masted boats as 'desirable' to American yachtsmen and British designers had to concede in the early 1960s that beam was not so disadvantageous to speed and windward ability. But because of her huge differentiation from the RORC theme of design, by now well established in Europe and Australia, it meant that it took a further 15 years to bring the CCA and RORC rules together and Olin Stephens, who uniquely had a run of distinguished designs to both rules, to do so.

A design which followed the same theme as *Finisterre*, but was less extreme if only because she was larger, was *Carina* designed by Philip Rhodes (USA) for Richard S. Nye. She won numerous races including the transatlantic races of 1957 and the Fastnet race of the same year and Class I in the 1959 Fastnet. Dimensions were LOA 53 ft 6 in, LWL 36 ft 3 in, beam 13 ft, draught 6 ft, displacement 32 000 lb, sail area 1200 ft². She was a yawl with a great variety of sails of the unbeatable shape and cloth that were being developed in the USA at the time. The 1959 Fastnet was in generally fine weather, but in 1957 there were continuous gales. As *Carina* crossed the finishing-line Dick Nye made one of the now most over-quoted remarks in yacht racing, 'OK boys, we're over now: let the damn boat sink.' All through the 1960s there were numerous boats in the USA with features of *Carina*, but none in Britain where two-masted rigs were rare and centreboards almost unknown in large yachts or ocean-racing boats.

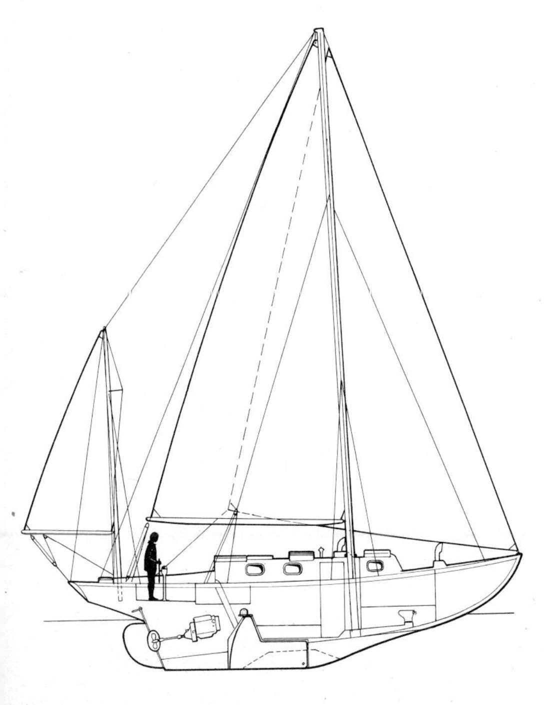

Profile and some detail of the yacht which had the greatest influence on offshore boat design in the early 1960s: *Finisterre* was a centreboard yawl of shallow draught, short masts and with ample accommodation and deck structure.

Right
Finisterre.

Large as IOR boats go, the 66 ft *Phantom* designed by Cuthbertson and Cassian, Toronto.

Right
Phantom on a road trailer.

The largest two-masted sailing-cruiser to be built in England since 1945 was *Blue Leopard*. The significance of the design is that it shows the departure of first-class cruisers from ocean-racing yacht design. The designs mentioned so far have been mainly ocean racers, because this is the area where development has been in sailing-craft since the second decade of the century. But this very development has made the yachts in some ways unsuitable for cruising. In *Blue Leopard*, the deck structure and ends, unaffected by rating rules, show her to be for grace and convenience. (If rated she would exceed the 70 ft top limit of the IOR.) Designed by Laurent Giles and Partners and built in 1963, the yawl rig had a mainsail of nearly 1000 ft^2, but she was a cruiser, in that with sails down, twin Rolls-Royce diesels of 240 bhp would drive her in any weather. Light displacement she owed to ocean-racing experience, long ends to the metre boats. Her dimensions are huge: LOA 113 ft, LWL 92 ft, beam 19 ft 3 in, draught 9 ft 8 in, displacement 115 000 lb.

The greatest change in the designed profile of the mass of sailing yachts since *Britannia* was from 1966 onwards when the rudder in new designs became invariably separated from the fin keel. Initially for racing boats under both CCA and RORC rules, it has spread to cruisers, 12-metres, and even the rules of the obsolescent 5·5-metre were changed to allow the arrangement. Though 1966 was by no means the first time the separate fin and rudder had been seen. It was seen on the raters of the late nineteenth century, it was on class boats like the Flying Fifteen, on earlier offshore racers by E. G. van de Stadt, Illingworth, and Laurent Giles.

But in the 1966 season in England the fin keel yachts *Roundabout* and *Clarionet* designed by Sparkman and Stephens swept all before them. Sister ships like *Al'nair* in Italy and the Cal 40 in America did the same abroad. *Roundabout* was a One Ton Cup boat (under the RORC Rule), owned by Sir Max Aitken – she was 36 ft 10 in, LWL 26 ft 8 in, beam 10 ft, draught 6 ft 3 in, displacement 12 400 lb. Olin Stephens had shown the way to make a hull form suit the 'fin and skeg' system with certainty. A year earlier the separate rudder boat *Rabbit* had won the 1965 Fastnet, she was 2 in over minimum size allowed for the race, with a waterline of

24 ft 2 in, LOA 33 ft 6 in, sail area 456 ft²; displacement 9950 lb. Other designers then followed these Americans in the profile of racing and then cruising yachts.

The largest sailing yachts to race regularly since 1945 have been the so-called 'maxi-ocean racers', which include *Southern Star*, *Ticonderoga*, *Windward Passage*, all American, and after them *Pen Duick VI* and *Great Britain II*. *Ticonderoga* was designed by L. Francis Herreshoff (son of Nathanael). The first boats were the product of the opulent 1960s, and the object of the American owners was to be first home regardless of rating. The CCA ruled a maximum overall length of 73 ft, so the designs became short ended, with as long as possible a waterline, and accepted a high rating in the interests of absolute speed. In *Windward Passage*'s first year, 1970, she was first to finish in every race in which she took part (but of course did not *win* all of them). Owned by Mark, and then his son Bob Johnson (USA) she is 72 ft 9 in, LWL 65 ft, beam 19 ft 4 in, draught 9 ft 6 in, displacement 80 000 lb, sail area 437 ft². It is a useful observation on modern yacht design to mention that on her keel is 34 000 lb of lead. There is an ample engine-room with a Westerbeke 4-107 diesel. Construction of the hull is triple diagonal spruce planking and she was built in Grand Bahama.

With the introduction of the IOR, it was seen that a length limit with no rating limit was not logical and the CCA came into line with the permanent maximum of 70 ft rating under the rule. The result is a 'maxi-ocean racer' with more graceful ends. In 1974, Sparkman and Stephens designed *Kialoa III* for John B. Kilroy of San Francisco. Constructed of aluminium she was built by Palmer Johnson of Sturgeon Bay, Wisc., and is ketch

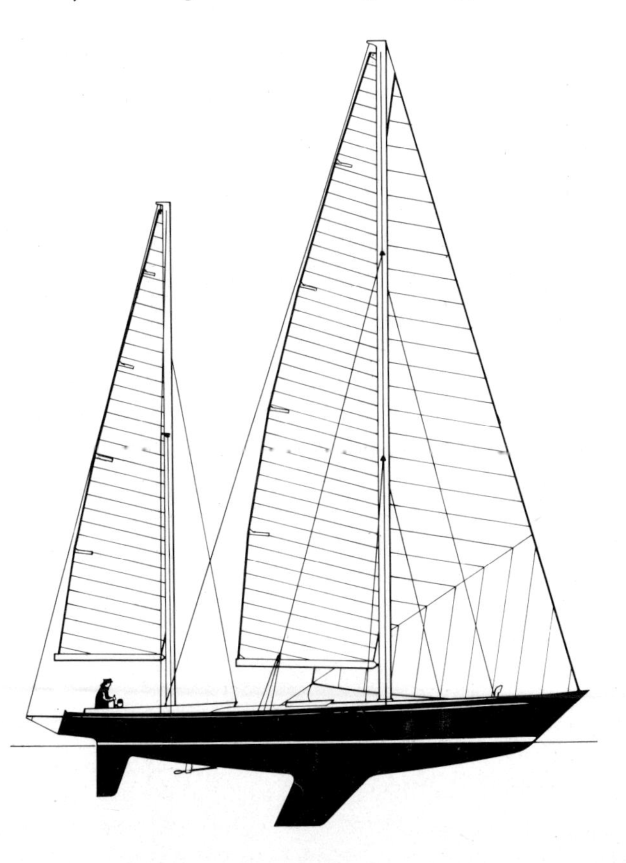

A prime example of the biggest size and type of yacht racing in 1975. *Kialoa*, American-owned 79 ft light alloy ocean-racing yacht near the maximum allowed by the IOR. Note minimum underwater profile, fin keel, rudder right aft, very tall narrow sails and big genoa, flush deck. There are permanent berths for ten crew, while 400 gallons of fuel and 600 gallons of water are carried.

rigged to keep individual sail areas manageable. She has LOA 79 ft, LWL 62 ft 6 in, beam 17 ft 5 in, draught 11 ft 7 in, displacement 84 440 lb, lead ballast 35 000 lb, sail area 2733 ft^2. Power is 74 hp Volvo MD 32A.

The most influential design in multihull development was the 70 ft trimaran *Manureva* (ex-*Pen Duick IV*). Prior to her design and success there was no set pattern for racing multihulls, but rather a mass of original ideas on different craft, both 'cats' and 'tris'. This was partly also because there was no rating rule to restrict or channel development. *Manureva* with her high speeds across the Atlantic, Pacific and round the world between 1969 and 1974 showed a certain line of design for sound multihulls. The emergence of an 'International Offshore Multihull Rule' in 1974 has yet to show any effect on design.

Gumboots, One Ton Cup winner 1974.

The typical style for sailing yachts of the mid-1970s is too close to the survey to be seen in focus. One American designer to emerge in the tradition of Burgess, Herreshoff, and Stephens is Doug Peterson of Pasadena, California. In the 1973 One Ton Cup his simple *Ganbare* was 2nd, but impressed with the best speed. Her successor *Gumboots*, the first GRP design to win the Cup, did so in 1974. In 1975 a boat of his design won the SORC. Other boats to his design did very well. The IOR sloop developed has a fin keel, short ends, high freeboard, minimum mainsail allowed under the rule. The section of *Gumboots* is remarkably different to the favoured offshore yacht of 1930 to 1965. The fin keel is very much a thin minimum area plate of streamlined pattern, the under body of the hull is flat. On deck and in the rig the boat is entirely designed for racing. Dimensions are LOA 35 ft 6 in, LWL 29 ft 6 in, beam 11 ft 5 in, draught 6 ft 4 in, displacement 14 000 lb; sail area 542 ft^2.

BUILDING BOATS

Boat industry

The largest boat-building and allied industry is, of course, in the United States of America, where that industry estimates that its customers comprise 47·2 million people 'who participate in recreational boating'. This is getting on for the whole population of Britain and is not surprising when the combination of affluence, the climate across the continent, and the thousands of miles of waterways, inland and coastal, protected bays, rivers and lakes, and the size of the population are considered. Some 700 000 families get their recreation in boats of less than 16 ft. The total number of boats in 1973 was estimated at 9·4 million with a growth rate of about 2½ per cent. Since then the energy and world situations have diminished the figure.

The greatest numbers of every type of recreational boat are also in use in the USA and comprise 40 000 sailing-boats, 745 000 inboard powered boats, 2·4 million rowing-boats, canoes, and small open boats, and 5·5 million outboard powered boats. The American trade associations are the National Association of Engine and Boat Manufacturers and the Boating Industry Association. In 1973 they reported a sale in retail outlets of boats, equipment, and on their maintenance of $42 000 000 000.

No other single country approaches the size of these statistics and this goes for the number of persons employed in serving recreational boating in the USA; 350 000 full-time employees and 100 000 part-time employees employed by 18 000 firms. Of these 2500 are in manufacturing and the remaining are retail dealers, distributors, and servicing concerns.

The second biggest national group of participants in recreational boating is in Britain with an estimated 2·75 million people, then come Australia and Germany, each 2 million; France 1·9 million; Italy 1·6 million, and the Scandinavian countries with 1·4 million. These are figures which the authorities in each country estimate and are no doubt very broad and very optimistic in what is meant by 'boating'. For instance the Italian scene, with numerous beach-type craft and casual power-boats, is quite different to the island holidaying Swedes, with their sailing boats in use during a short summer. However, even adding these totals together with other European countries does not reach the United States total.

The great increase in the popularity of the sport has gone hand in hand with the production of easily maintained and designed and built GRP boats. The cause and effect are mixed. In simple figures the increase in consumption of glass-reinforced plastics in boat-building is remarkable. For the main countries it looks like this:

TONNAGE OF GRP

	1961	**1968**	**1970**
USA	25 000	72 000	118 000
Britain	2000	8400	13 400
France	900	7300	11 000
Germany	425	3700	5500

The greatest annual sales of boats, engines, sails, and all equipment and including hire fees and export sales from Britain was in the year ending 31 March 1973. The turnover value was £100 352 000 ($238 837 000). This was reported by the trade federation for British boat-builders and linked trades which is the Ship and Boat Builders' National Federation. It also assessed the number of boats plying British waters in 1973 as 575 000. The largest proportion of these, just under half, were small open rowing-boats, dinghies, canoes, and so on. Just under one-quarter were class sailing dinghies of all materials and some built with do-it-yourself kits. The rest were small open power-craft (15 per cent); speedboats (8 per cent); seagoing sailing-yachts (9 per cent), and motor-cruisers with cabins (3 per cent). An estimated 175 000 outboard motors are in use in Britain. (Unlike motor cars, of course, they spend most of their time hanging on a rack or on the stern of a moored boat doing nothing, but this could be said to apply to most pleasure-boats items.)

The number of member firms of the Ship and Boat Builders' National Federation in 1973 was 926 (or just over 5 per cent of the number in the USA). Most British boat-building firms, despite steady take-overs, are small. Most of these firms had a turnover of less than £100 000: 389 firms even had a turnover of less than £25 000! Eighty-nine firms turned over more than £250 000 and just under half of these over £500 000 million pounds.

The works of the Morgan Yacht Corporation.

The builders of the greatest numbers of sailing boats are three American firms, each of which is owned by a parent 'conglomerate'. They are Cal-boats of Costa Mesa, California, which is owned by Bangor-Punta; Columbia Yachts, of Chesapeake, Virginia, owned by Whittaker Corporation; and Morgan Yacht Corporation of Fort Lauderdale, Florida owned by the Beatrice Group. In 1973 Morgan was building 1000 yachts per year of a size between 27 and 41 ft.

The oldest boat builder still trading in the same business is Camper and Nicholsons of Gosport and Southampton, England. The firm was owned by the family of Nicholson until 1972 when it was bought by Crest Proper-

Part of Camper and Nicholsons' yard at Gosport. On the left is Edward Heath's fourth *Morning Cloud* which had just been launched on this day.

ties of London. The members of the Nicholson family remained in active management. In 1780 a Mr Amos was known to have a boatyard at Gosport, near the entrance to Portsmouth Harbour (at that time the Navy was very active in the port although the Battle of Trafalgar was yet 25 years off). In 1809, Master Camper was apprenticed to Francis Calenso Amos and in 1842 Benjamin Nicholson was apprenticed to Camper. By 1855 Camper and Nicholson were in partnership and the firm took that name. The present managing director and successful racing yachtsman, Peter Nicholson, is the great grandson of Benjamin. Benjamin's son, Charles E. Nicholson, was the designer and builder of *Shamrock V*, *Endeavour I* and *Endeavour II* (see p. 136) as well as numerous other yachts. Between 1896 and 1925, Camper and Nicholsons built (excluding day boats and one-designs) 224 yachts, totalling 14 704 tons; between 1925 and 1960 there were 400. These included the largest ever from the yard, the motor-yacht *Philante*, 1629 tons – now the Norwegian royal yacht *Norge*. Also in this period, the yard built 4 J-class yachts, 4 23-metres, 1 19-metres, 3 15-metres, 16 12-metres, 15 8-metres, and 18 6-metres. Two sailing yachts to have been owned by the Queen and Prince Philip, *Bloodhound* and the Dragon class *Bluebottle* were built by Camper and Nicholsons and they are now the only yacht-builders 'by appointment' to the present monarch.

After 1960 many hulls of glass fibre were finished between 30 and 53 ft at the rate of approximately 100 per year.

The first motor made by Ole Evinrude in 1909.

The unsurpassed explosion in yachting, especially after about 1960 is emphasized by such figures and their comparison to those mentioned above for the production of Morgan. Up to 1925, therefore, Camper and Nicholsons were building 7·5 yachts per year of average 65·5 tons, say 100 ft length (displacements were heavier then than now). Then in the next 35 years, 11 per year average, and after that the 100 mentioned, but of average length 38 ft. As Camper and Nicholsons is considered at the more refined end of the yachting market, most builders in England would produce a smaller size average.

The biggest production GRP yacht in the world is the Ocean 71 built by Southern Ocean Shipyard of Poole, Dorset, England. It has held this record since the first one was produced in 1970. The yacht is a ketch of 71 ft designed by E. G. van de Stadt of Zaandam, Holland. Larger GRP vessels have been built, notably minesweepers for navies. These are up to 125 ft long, but are not 'production boats', each one being designed for her purpose.

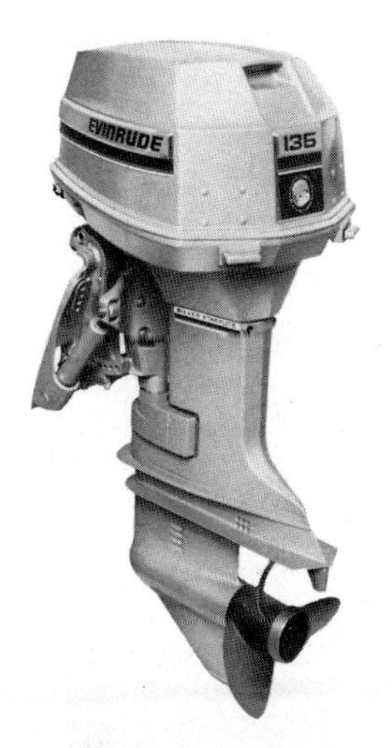

A 1975 Evinrude, 135 horse-power with power trim and tilt, 12 amp alternator and highly developed propeller.

The first class yacht to be production moulded in GRP was the Flying Fifteen in 1952 by Halmatic Ltd of Portsmouth.

The first outboard motor was invented in 1903 by Cameron B. Waterman of Detroit, when a student of Yale Law School. It is said that he thought of the idea when slinging a bicycle motor over the back of a chair. Waterman began manufacturing them in 1905 and sold 3000 a year, which was below his expectations. In 1916 the Waterman business was sold to Arrow Motor and Marine Co., of New York, which itself went out of business in 1924.

The first successful manufacturer of outboard motors was a Norwegian-American Ole Evinrude, who sold the first ten hand-built motors in 1909 at $62 each. Sales were slow until the introduction of the Elto (Evinrude Light Twin Outboard) in 1920. By 1929, 5500 motors per year were selling in four major makes in the USA. Evinrude motors later merged into the present Outboard Motor Corporation (OMC).

The most complete list of yachts available in Britain is *Lloyd's Register of Yachts.* It is published annually and gives the owners, main dimensions and similar particulars, and whether the yacht is classified for sound construction under the various rules of *Lloyd's Register* for different materials. It was the first list to be published anywhere and the first edition came out in 1878. The *Register* covered, and still does, yachts in countries other than North America where since 1903 there has been published *Lloyd's Register of American Yachts.* Apart from club lists and individual attempts to combine these, Lloyd's has always been unique. Neither Lloyd's nor any other list is comprehensive because it relies on voluntary supply of information by owners, designers, and yards. To keep the work within bounds there is a lower limit for size of 250 ft^2 sail area or 30 ft for motor craft, so numerous craft scattered all over the coasts are not shown. Many large craft (see p. 216) owned by companies of one sort or another which are registered (both company and yacht) in countries where there is available a 'flag of convenience', have no reason to publicize themselves by appearing in *Lloyd's Register.* Lloyd's does contain some lists of small craft which belong to international racing classes, giving owners, sail numbers, and brief class details. Classes like Dragons, Stars, and so on are included, though below minimum size for individual registry.

A clarifying note about registering and Lloyd's

In different countries there are varying rules about the official registration of yachts. In the USA, States have their own rules and in many cases boats with motors over a certain horsepower have to show numbers on their hulls. In Britain there is no compulsory system of registration. Most boats over about 25 ft will register as a British ship with the Department of Trade: and there is a register of ships, which includes yachts, which can be seen at certain Government offices, but is nothing to do with *Lloyd's Register.* Lloyd's is an independent organization and its surveyors supervise the construction of yachts (and ships) and then issue certificates to state that

The marina at La Rochelle

the vessel is in accordance with their rules (such as '+100 A1'). Yachts in their register may or may not be Lloyd's supervised, but if they are, this is indicated. The famous insurance organization, Lloyd's of London (which does much shipping and yacht insurance) is yet another quite separate body.

The oldest wooden yacht with full Lloyd's classification (i.e. +100 A1) is the 47 ft 6 in yawl *Saunterer* (beam 10 ft 4 in), designed and built by Charles Sibbick of Cowes in 1900. The classification was confined as being in order in 1974 and since 1959 the yacht has been under the ownership of Alexander Sutherland and based on the Clyde. The yacht was owned by Captain Oates of Scott's 1912 Antarctic Expedition from 1906 until his death.

Yacht harbours

The greatest problem which the expansion of yachting has brought to the owner of a new boat is finding a berth in which to keep her. Centreboarders and open boats can be hauled out on dry land and taken home on a trailer or left in a 'dinghy park'. Larger boats need a berth with ample water at all states of tide and reasonable access. All free-swinging moorings and piles and buoys in rivers and similar sheltered channels were long ago fully occupied.

The modern solution is to build a special yacht harbour or marina by dredging out mud or sheltering previously exposed water. Marinas originated in Florida and California; yachts were packed in by having a network of floating 'docks' (USA) or 'pontoons' (Brit.). The word 'marina' is often disliked in Britain and the term 'yacht harbour' preferred.

The biggest marina in the world is the Marina del Ray in Los Angeles, California. There are berths for 7500 yachts. The second biggest has berths for 3000 yachts at Alamedos, California.

The biggest marina in Europe is the one completed at La Rochelle, France in 1974, with berths for 2500 craft.

Saunterer with her modern rig on the Clyde.

Right
Saunterer with her original rig in 1900.

The biggest in Britain, when completed in 1977, will be Brighton Marina. It will have berths for 2313 craft. It is also claimed to be the biggest yacht harbour in Europe under single management. Built in the open English Channel, it is in effect a new port on the coast of England, where only exposed beach existed previously – 126 acres will be enclosed by it, with 77 acres for moorings. It is 1000 metres across and 550 metres from the offshore boundary to the cliff which backs it. The maximum draught in the tidal basin will be 11 ft. In the complex there will be parking for 4600 motor cars. The outer arms of the yacht harbour are being constructed by sinking pre-cast concrete caissons of 40 ft diameter and weighing up to 625 tons each – 110 of these are being used.

The biggest yacht harbour in use in Britain in 1975 was the Chichester Yacht Basin in Sussex. It is landlocked with berths for 900 yachts and access is by lock when the tide outside is suitable.

The number of marina berths in Britain in 1975 was 8500 in 42 existing marinas. Thousands of other boats were on various other sorts of mooring in creeks and estuaries, in mud berths or hauled out.

Large motor yachts

The largest and most expensive to build and maintain of all yachts are motor yachts which may be 200 ft or more. They merge into the realm of private passenger vessels rather than the 'yachts' referred to in most of these pages. It is not possible to say precisely where private pleasure vessels end and yachts which belong in effect to the navy of a country begin. The latter might be royal or presidential yachts. The British royal yacht *Britannia* comes into such a category. The question also arises as to what yachts can be said to be 'private'. When one begins to look at such craft, they are very often found to be owned by companies registered in various parts of the world. These craft are intended to be elusive by their rich owners. Their existence is of a totally international nature. The flags with the most tonnage of large motor yachts under them are Panamanian, Liberian, and Japanese. Seldom will a yacht under such a flag be in the country. Very likely she will be in a small port in the Greek Islands; the crew may well be British (much prized especially if the Captain and Engineer) or Italian. A charter party may be flying to join her for several weeks from New York; the person with the controlling interest in the owning company may be on business in South Africa.

However, this system of management is a product of the world since about 1960, for Governments everywhere levy elaborate taxes and international currency is a complexity. Before then there were a number of rich individual owners of fine vessels.

The first steam yacht was built in 1829 and owned by Thomas Assheton-Smith and was *Menai* (400 tons). Assheton-Smith had to resign from the Royal Yacht Squadron to own the vessel. He also had seven others built subsequently.

The only attempt to stop the spread of the steam yacht was the reason that Assheton-Smith resigned from the Squadron. In 1827 at a meeting at the Thatched House Tavern in St James's the following Resolution was passed: 'Resolved that as a material object of the club is to promote seamanship and the improvements of sailing vessels, to which the application of steam engines is inimical, no vessel propelled by steam shall be admitted into the Club and any member applying a steam engine to his yacht shall be disqualified thereby and cease to be a member.' However, in 1844 the rule was changed to allow members to have steam yachts, if over 100 horsepower. In 1853 all restrictions on motor craft were removed.

The first steam yacht in the United States was *North Star*. She was built in 1853 and owned by the richest private citizen in the world, Cornelius

Vanderbilt. *North Star* was 270 ft and had side wheel paddles of 34 ft diameter. On her maiden voyage with the owner and his family, she left her building yard in Greenpoint, New York, and visited many countries in Europe making a voyage of 15 024 miles. Her best speed over an hour was 14·5 knots. On her return to the USA she became a passenger steamer.

The largest private yacht ever built was the 4600 ton steam yacht *Savarona*. This was constructed as late as 1931 for Mrs Emily Roebling Cadwalader (US) an heiress and grand-daughter of the builder of Brooklyn Bridge, by Blohm and Voss of Hamburg and designed by the US firm of Gibbs and Cox. In 1938 she was sold to Turkey and is today a school ship in the Turkish Navy. *Savarona* was 408 ft 6 in and carried the largest crew of any steam yacht – 107. The yacht was only in commission for two seasons and was never in the country of her owner, to avoid taxation.

The second largest private (as opposed to royal, presidential, or naval) yacht was built to the order to J. P. Morgan, the American banker. The yacht *Corsair*, was the fourth of the name designed by H. J. Gielow and built by Bath Iron Works, Maine. Length was 343 ft 6 in and tonnage 2142 gross. There were two turbo-electric engines. Her maiden voyage took her from Long Island Sound to Cowes, England, in 7 days 7 hours. *Corsair* was launched in 1930 and because of the economic depression, the public was excluded. Being America a tugboat was, however, provided for the press! Morgan gave the yacht to the Royal Navy in 1940. After the war, she was refitted and used as a cruise ship commercially off the Pacific coast of North America and was wrecked off Acapulco on 12 November 1949.

The most expensive yacht for fuel against mileage was *Valfreya*, 150 ft. Owned by the recluse American millionaire, M. Bayard Brown, in 1865, the yacht never left her mooring, but Brown ordered steam to be kept up continuously in case he might want to get under way at a moment's notice. He never did and *Valfreya* was later sold to the Maharajah of Nawanger and became the *Star of India*.

The first diesel-engined yacht of any size and which was habitable with sleeping cabins was *Pioneer*, designed and built by Camper and Nicholsons in 1913. She was 163 ft long and owned by Paris Singer: the diesels were 250 hp Atlas Polars built in Denmark. The advantage of diesel over steam was that the engines took up much less space; diesel oil was also cheaper (though it might be wondered if this mattered to such rich owners). The arrival of *Pioneer* could be said to coincide with the peak of the steam yacht's period, for after the First World War, few were built – diesel engines became the norm. Even in 1954 Aristotle Onassis bought a steam yacht, the oil burning reciprocating engined ex-anti-submarine frigate, which was converted to the yacht *Christina*, 325 ft. *Christina* and another yacht *Moineau* converted in the same way were the only two new steam yachts after 1945. (In 1973 *Moineau* was for sale in the Caribbean for $1 000 000. *Christina* remained with Onassis.)

In 1913 the distribution of steam yachts by country was as follows: USA 272, Britain 263, France 43, Canada 22, Russia 12, Germany 9, other countries 84. Never again, in steam, was this total of 696 reached.

The largest private motor yacht in the world today is *Atlantis* of which the beneficial owner is the Greek shipowner Stavros Niarchos. Dimensions are 332 ft 3 in, beam 47 ft 2 in, draught 19 ft 8 in, 2585 tons gross. *Atlantis* was built by the shipyard of Schlicting at Travemunde, Germany and designed

Atlantis, used by Stavros Niarchos, 332 ft.

by Maierform. There is a crew of 50 and facilities include a cinema and swimming-pool. Paintings by Van Gogh, Gaugin, and Renoir have been hung in the yacht. The hull is steel with two Pielstick oil engines.

The largest yacht in the world today that is for charter is not a motor-yacht but the four-masted barquentine *Antarna*. She is square-rigged on three masts, the main having six square sails. There are four Enterprise diesel engines giving over 15 knots which will get her out of any sailing problems. There is an American crew of 50. With LOA 316 ft, beam 49 ft 2 in, *Antarna* was launched in 1931 as *Hussar* in Germany for Edward F. Hutton; later the vessel was named *Sea Cloud* and owned by his daughter, the millionairess Barbara Hutton. An American company now owns her.

Antarna, four-masted barquentine yacht.

The number of large motor-yachts today in the world is to some extent indicated by the number available for charter at any time. Such charters are, of course, not simple hiring, but a matter for negotiation between the charter broker and charterer and presuming the yacht is not required by the owner at that time. In 1975 the leading London yacht charterer, David Halsey, had available in various parts of the world, 87 yachts of between 100 and 200 ft long and 4 yachts of over 200 ft.

The largest charter fee quoted between unnamed clients of a yacht charter broker is $4000 per day. Such yachts are often chartered to foreign Governments, film companies, research expeditions, and other organizations.

The largest yacht for occasional charter of British build and with a British crew of 30 is *Shemara* – LOA 212 ft, beam 30 ft. The yacht was built in 1938

Shemara, used by Harry Hyams.

by J. I. Thorneycroft, Southampton and is powered by Polar Atlas diesel engines. For long owned by Sir Bernard and Lady Docker, who used the yacht with considerable publicity, the beneficial owner is now Harry Hyams, British property developer and financier.

The modern German motor-yacht *Carinthia VI*, designed by Jon Bannenberg.

The most modern design of large motor yacht is *Carinthia VI*, designed by Jon Bannenberg of London for the German stores magnate, Helmut Horten. Her dimensions are LOA 223 ft 10 in, beam 29 ft 10 in, 874 tons gross. The yacht was built of steel by Lurssen Weft, Bremen, Germany with triple screws to drive her. Her home port is Cannes.

Multihull yachts

The most radical design change to make an impact in recent years in sailing is the multihull yacht. Its origin is old but until about 1950 the concept was rare in pleasure sailing. The first multihull boats were catamarans in the Pacific: the word 'catamaran' in ancient Tamil means 'tied trees' and referred to logs tied together. The word today means any two-hulled craft, power or sail, and also refers to rafts like buffers used between the side of big ships and a wharf. Three-hulled craft are now distinguished by the term 'trimaran', said to have been coined by Victor Tchetchet (Russian-American). Catamarans of ancient type were used by Melanesians, Micronesions, Indonesians, Singhalese, and Polynesians. The last are the best-known seafarers, and are believed to have had craft up to 100 ft long which carried up to 400 persons at a time across the Pacific. When Western navigators saw the smaller multihull craft of the Pacific in the eighteenth century, they did not appear to be of great application, as they were not readily adaptable to carrying cargo and guns. It was these considerations which meant there was no commercial development in Europe of multihulls. Remember that European sailing ships depended for correct stability on the structure and cargo stowage, while the capacity of a multihull is critical.

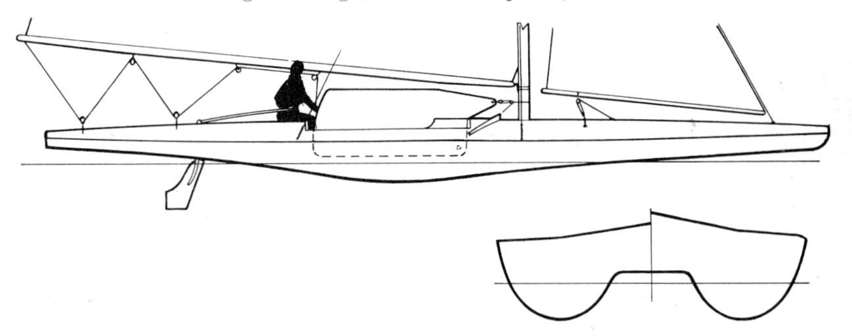

The Canadian designed and built *Dominion* which challenged for the Seawanhaka Cup in 1898. Because the Seawanhaka Rule took into account length and sail area it was possible to cheat the rule by building a twin-hulled boat.

There were isolated experiments with multihulls in the course of yachting. The first yacht race involving a multihull was in 1661. A catamaran had been designed as an experiment for Sir William Petty of Dublin. The boat apparently consisted of two closed cylinders 15 ft long and 4 ft 6 in apart. There were two rudders and two masts carrying 600 ft^2 of sail. It was built of wood. The catamaran raced several single-hulled sailing vessels: 'the King's open barge, Captain Darcy's Man o' War boat and several others. Sir William led to a mark ship several miles downwind in the Irish Channel. When rounding the light ship Sir William's nearest competitor, Captain Darcy, was quarter of a mile astern. Captain Darcy was fully confident that the next leg to windward would settle the matter; however Sir William sped away to finish one mile ahead.'

In 1898 G. Herrick Duggan (Can.) designed and built the 35 ft 10 in catamaran *Dominion* to challenge for the Seawanhaka Cup. The Cup was raced under the Seawanhaka length and sail area rule: and the 'cat' was therefore a rule-cheater. Its rivals were single-hulled scows, which above the water-line did not look very different. Any modern rule for single-hulled yachts would reject a twin hull. Nathanael Herreshoff also designed several sailing catamarans on scow lines: they were of a light and experimental type.

The first catamaran to be sailed offshore for any distance by a European and, therefore, the first offshore multihull as the term is intended today (as opposed to unreported Polynesian voyages), was *Kaimiloa*, 38 ft. She was built by Éric de Bisschop (Fr.) in Hawaii, who with Joseph Tatiboet in 1937–8 sailed the yacht from there via the Cape of Good Hope to Cannes, southern France. The quaint schooner-rigged *Non Pareil* in 1868 sailed from New York to Southampton in 43 days and this might be called a multihull. It is better described as a raft, which blew with the prevailing wind.

First yachting trimaran, designed and built by the Russo-American Victor Techetchet in 1945.

The first trimaran to enter the yachting scene was a 24 ft day racer and was more in the nature of a conventional though narrow dinghy hull with small floats on each side. It was designed and sailed in 1945 in Long Island Sound by Victor Tchetchet, a Russian-American. The first trimaran to sail distance offshore was the 43 ft *Ananda* from Cape Verde to Martinique in 1946. The skipper was the builder, André Sadrin, and he had a crew of two with him. There was no general interest among yachtsmen in the possibilities of the trimaran until about 1960, when Arthur Piver (US), who had developed a certain style of trimaran, sailed *Nimble* from Swansea, Mass. to England. He had hoped to be in time for the single-handed transatlantic race, but was late. Piver was not slow to seek publicity and although he made sure that by then all yachtsmen had heard about 'tris', his extravagant claims for speed had a backlash and trimarans and multihulls generally were considered dubious in many ways. First the speed claims were disbelieved and the light build of the Piver type was easily damaged by driftwood and when manœuvring in harbour.

The first ocean crossing by a catamaran after 1945 was by the 47 ft 8 in *Copula* from Bordeaux to New York via Florida and the West Indies in 1950. The builder/owner was Raoul Christiaen (Fr.) and she was built of steel and was heavy (and slow). She was even narrow for a multihull, 17 ft 9 in. The low-aspect Chinese junk sails made sure there were no stability problems. She was not a type that was copied, although she was the largest catamaran at the time of her voyage, with a crew of five.

A large proportion of interest in multihulls throughout the world was occasioned by the regular publications from England of the Amateur Yacht Research Society. All sorts of craft were described and discussed in these booklets which appeared regularly from 1955 onwards.

A reminder – Why more than one hull?

It has just been mentioned that the single hull developed in Europe, because for cargo-carrying capacity the single hull could be loaded to give stability under sail. As yachting developed, so the keel as we now know it enabled a boat to stand up to her canvas. The keel boat can be tacked quickly and manœuvred easily in narrow spaces, she can come alongside and small ones can be towed on a road trailer. The snag is that the price paid for stiffness under sail is actual weight (of the lead or iron keel) which has to be dragged through the water. In order to push displacement (weight) through the water, more sail is needed. In practice in the efficient deep-keeled yacht, a balance is struck between displacement, length, and sail area.

The centreboard dinghy breaks out of this mildly vicious circle by using the crew as ballast. The members of the crew hang over the side, perhaps on trapezes to resist the heeling power of the sail area. So the boat is light and *planes*. The secret is in leverage of the main weight on board (the crew) and its adjustment to the windward side on each tack. But this cannot be done on large craft (over 20 ft say) because the actual human weight is not great enough. Of course, the crew sit out on big yachts and this aids the power to carry sail for racing: it is not the main antidote to heeling.

The multihull gets rid of ballast on large boats by depending for its stability on the great beam of the hull. If it was *just* a wide hull, like a barge, there would be too much friction of the water and hull shape resistance. But perched on two or three finely shaped hulls there is lack of skin (water) friction and the widely spaced hulls prevent heeling: so the power of the sails is transmitted into forward motion Best of all, that weight has gone. There is no lead to carry about. The result of a light hull but ample sail is speed and this is obtained from well-designed multihulls (as *Pen Duick IV*). Badly designed ones or intentional cruising 'caravans' are even slower than single-hulled yachts because they are undercanvased for fear of tipping. Tipping is the danger which remains with multihulls and is not generally present with keeled yachts. Multihull stability comes from hull form, but if the angle of heel is beyond say 40 degrees this quality of the hull is lost. Over they will go and then take up the highly dangerous situation of having stability upside-down. A keeled yacht, on the other hand, will bob back up because of keel weight, though if she *fills*, due to structural damage letting in water for instance, she will sink. A multihull turned over will continue to float: there is no lump of lead to carry her to the bottom. This is not much compensation for capsizing miles from land. Despite many multihull voyages, a single-hulled keel vessel would remain the choice of the vast majority of cruising men (95 per cent according to a *Yachting World/Sail* survey in January 1973).

The other disadvantages of multihulls include space required in harbours and at moorings, manœuvring in close quarters and for coastal cruising sheer speed is frequently not the main requirement. Advantages besides speed are good deck space for charter parties and families and the comfort obtained from small angles of heel in all ordinary wind strengths. To obtain the advantages of a multihull there is a minimum size of design. When the yacht is below about 35 ft the scale effects of naval architecture make both stability and living space difficult to obtain; then a single-hulled cruiser has most of the advantages. Small multihulls in effect become the equivalent of centreboard dinghies and high speeds can be obtained for thrills and racing.

An early Shearwater catamaran.

What has really made possible the optional selection of a type of hull, trimaran, catamaran, or single hull, is the introduction of modern materials, especially GRP. With this and aluminium (for struts between the hulls) it has been possible to get the hulls the desired shape and of the needed strength. In the days of wood this was not feasible, except for the occasional one-off experiment as we have seen.

The first catamaran dinghy class to race regularly was the 16 ft 6 in Shearwater in 1958. There are now 1600 in the world. The design was by Prout Brothers of Canvey Island, Essex. The first multihull to enter the Olympic Games is the *Tornado* in the 1976 Games.

Blue Leopard, designed by Laurent Giles and Partners, represents the finest type of modern cruising yacht: high performance under sail, high performance under power, comfort below, elegance by any standard

This ocean racer to the IOR represents the latest thinking in design by 1974 when she was launched in Holland to an Australian design for an English owner. *More Opposition*, 49 ft, designed by Bob Miller: the yacht was a 1975 Admiral's Cup contender

The largest production glass-fibre sailing-yacht in the world is the Ocean 71 class designed by Ricus van de Stadt and built in England

The largest standard production sailing multihull in Britain is the Solaris cruising catamaran. Built in GRP with considerable accommodation across the bridge deck, has enabled a saloon and *four* double berths to be fitted, The dimensions are LOA 41 ft 10 in, beam 17 ft 9 in, draught 3 ft 6 in, sail area 675 ft^2, displacement 16 800 lb. This design was by Terry Compton (Brit.) and the cost was about £35 000 in 1975. There are larger catamarans in various parts of the world, but they are one-off. An example of this is *British Oxygen* (p. 54) which is built for racing by a small crew. Apart from looking quite different to Solaris there is a complete change in concept in that the two hulls on *Oxygen* contain the accommodation and the hulls are held together by minimum structure: on Solaris the area between the hulls is used for lavish cruising accommodation. This highlights a way in which multihulls differ from single-hulled sailing yachts. The multis have not been in modern development for very long (25 years) and there has not been any national or international rating rule to push designers along certain lines. Individuals have, therefore, had a free hand and the appearance of the multihull yacht varies very greatly.

The largest standard production trimaran is the Cross 50 designed by Norman A. Cross (US). Built of GRP, LOA is 49 ft 9 in, beam 27 ft, displacement 24 000 lb. There is a yawl rig and 55 hp inboard engine. The largest production trimaran in Britain is the 34 ft 9 in Rapier, designed by Andrew Simpson and built by Simpson Wild Marine of Swanage, Dorset. Other dimensions are beam 24 ft, draught 1 ft 7 in, and 5 ft with dagger board through centre hull. The outside hulls are dismountable for transporting by road.

The largest sailing multihull is the foam sandwich trimaran *Great Britain III* designed by Derek Kelsall for Chay Blyth to compete in the 1975–6 Atlantic criss-cross race for multihull yachts. The dimensions are LOA 80 ft, LWL 70 ft, beam 38 ft, draught 2 ft 7 in (with centreboard 11 ft 10 in), displacement 30 900 lb.

The largest trimaran and catamaran prior to the Chay Blyth boat were *Manureva* (ex-*PenDuick IV*), 70 ft, and *British Oxygen*, 70 ft, designed by Rod Macalpine-Downie.

Chay Blyth and Jack Hayward plan *Great Britain III*, the world's largest trimaran.

A hull of the world's largest trimaran taking shape.

Multihull design is a specialist field and is not often mixed with single-hulled sailing craft. In Britain the designers of multihulls include Terry Compton, Chris Hammond, Derek Kelsall (whose *Toria*, 42 ft, which won the 1966 Round Britain race, inspired Eric Tabarly to build *Pen Duick IV*), Tom Lack, Rod Macalpine-Downie, Robin Musters, R. and F. Prout, Andrew Simpson, James Wharram, and John Westell. American designers include Dick Newick (designer of particularly fast trimarans such as *Gulf Streamer* and *Three Cheers*), Norman A. Cross, Rudy Choy (specializing in fast downwind catamarans for the Pacific coast and crossings), Ed Horstmann, and Lock Crowther, to whose design in 1974 was built the American trimaran with the biggest sail area to date, *Spirit of America*, with 2400 ft² on a 12 000 lb displacement. Arthur Piver, the US pioneer trimaran designer disappeared at sea in 1968 when on a single-handed passage down the Californian coast in one of his own boats, as did designer Hedley Nicoll, his contemporary in Australia, with a crew of two in a typhoon in the Pacific in 1967.

Construction

The basic materials of boat construction are more amenable to variety in materials than other vehicles. Wood, metals, and plastics are used widely: modern construction methods mean that it is often not possible to identify of what material a boat has been built without close inspection. These three basic materials have distinct variations in the way they are used.

The strongest method of construction is in the use of metal; if steel is used there are special high-tensile steels that enable a thin skin to be used. This might go down to 3 mm in the smallest type of yacht. Though steel has the best strength and stiffness, the disadvantages are that it is heavy, subject to corrosion, and difficult to insulate. Many alloys have, of course, been used in shipbuilding. Particularly famous in construction of yachts was the 1893 defender of America's Cup, *Vigilant*, which had steel topsides and a tobin bronze bottom. This worked all right for a short sailing season with the yacht frequently hauled out. However, a vessel built in the Far East in such a way was said to have been successfully beached just before the bottom and topsides separated! High-class racing yachts in the early part of this century did, however, use the system. Development of anti-fouling paints made it possible in the 1930s to have good yachts again built completely of steel. The last America's Cup series in J class yachts saw both yachts built of steel. Then the Cup was switched to 12-metres, the competition reverted to wood because this was essential to the rules of the 12-metre class.

The lightest material for its strength is light alloy – aluminium. This is now the material used by first-class racing yachts and also by 12-metres since 1974, when aluminium was first allowed in that class. The lightness can be demonstrated by the tremendous saving of weight of an aluminium 12-metre against a wooden one. It is 40 per cent. The growth in the popularity of aluminium is shown by the percentage of contenders for the three-yacht British Admiral's Cup team. In 1971 3·7 per cent of the contending yachts were aluminium; in 1973 27 per cent; in 1975 60 per cent. Modern techniques have improved the fairing and finish of aluminium boats and the 'crinkles' which were seen in earlier years are no longer a feature. The main disadvantage is the very high cost of building a boat in this material. In wooden and steel boats aluminium is popular for structures and as early as 1937 the British ocean racer *Erivale* had an aluminium coach roof. Aluminium is, of course, standard for masts and spars, as discussed below.

Building a boat of cold moulded ply using modern glues.

Wood is the most traditional material for yacht building, but rapidly becoming the rarest. In the London Boat Show of 1959, 69 per cent of the boats were wood. In the 1975 show 4 per cent were wood. Since time immemorial vessels have been built by placing plank upon plank: when these are edge to edge fastened to frames (known as 'timbers') this is 'carvel' building. Experience has perfected such a method, but craftsmen are becoming exceedingly rare throughout the world. In 1945 there were hundreds of small yards that would build yachts, boats, and launches in this way. Now there are but a handful that can build a yacht like this, although there are a number of small boat-building yards that can cope with open boats and launches.

Variations of planking include clinker or lap-strake, where the planks overlap; strip, where narrow planks are glued together; and double diagonal, where thin strips are laid at angles to each other. The advantages of wood are that a fair hull can be obtained through sanding and finish, but the finish is pleasing with good insulation, strength is good when built by a really first-class yard under proper supervision, repairs are comparatively easy. The disadvantages are ever-increasing cost due to scarcity of skilled labour, subjections to rot of various sorts and local structural defects.

The strongest method of building with wood is by using modern synthetic resin adhesives. These involve the processes known as 'cold moulding' or 'hot moulding'. In the latter thin veneers are inserted over a pre-tailored hot bag and the wood cooked with the glue. Much more common is cold moulding where thin veneers of boat-building wood are glued strip upon strip using a special resorcinol glue. The result is that the entire hull is a piece of plywood regularly shaped. The advantages are lightness and stiffness, but the wood can still be subject to rot. Construction, particularly in regard to the critical matter of glueing, must be undertaken by a very experienced yard under supervision.

Some small boats are still built of sheet plywood and this is most popular for do-it-yourself kits. Sheets of plywood are cut to shape and glued, screwed, and sometimes stitched, as in the Mirror dinghy. The advantages are cheapness and ease of construction. The disadvantages are that it is difficult to achieve the ideal shape and that the wood, once again, can deteriorate, which can include delamination even on the special waterproof plywoods used.

The most popular method of construction today is the use of glass-reinforced plastics. An advantage is that labour is easy to train to lay up the resin and glass into a prearranged female mould. When the boat is released it immediately gives a hull of exactly the required dimensions with a perfect finish. All other boat-building processes can then follow on it. It is completely immune to insects and borers; it cannot corrode; the weathering and degradation due to immersion in water are very slow. There is considerable resistance to impact and there is flexibility. The last-named can be a disadvantage for all types of boats where rigidity is required for performance and to prevent fatiguing and leaks. However, this is normally overcome by use of adequate bulkheads and structural arrangements. Glass-fibre boats were first constructed only in 1942: however since hulls have lasted since then life of a glass-fibre hull is not known, but with proper initial construction and reasonable maintenance it is expected to be in the region of 40–50 years. Cost is low where a good production run can be ensured, but would be high for a single boat because of the manufacture of plugs and then moulds.

A favoured system using glass and resin where only a single boat is to be built is to use foam sandwich construction. Here a glass-fibre skin is put over a male plug and then foam applied after which the outer shell is attached. The advantages of this are that the female mould is not required and considerable rigidity is obtained because of the two skins with the ridge of foam in the middle. In this sandwich construction one of the main difficulties is fairing the outside of the hull. Boats of this material have made long voyages and it is popular for big vessels that are intended to be light. Difficulties arise if the outer skin is damaged because of leakage between the skins and in some examples the thin skins have failed to hold their shape causing wrinkling.

The most misunderstood method of construction is the use of ferro-cement. This has found particular favour with cruising boats where one or two men wish to build it themselves. A wire and mesh framework to the desired shape is built round frames. The mortar is then mixed and applied to this structure very thoroughly to give a smooth hull and then cured under controlled conditions. After that the hull is fitted out in the ordinary way. The type of cement used is ordinary Portland or sulphate-resisting cement. The outside is smoothed by plastering or trowelling. The deck may be also of ferro-cement or may be of wood on wooden deck beams bolted to the edges of the shell. The advantages are that the boat can be built cheaply and special moulds and expensive materials are not required. The framework is tedious to build, but it is incorporated in the actual hull and, therefore, the builder can ensure that he is getting the exact boat he wants before applying the cement. The disadvantage for most yachts of this construction is weight, but techniques have improved and a boat of this material actually won the Sydney to Hobart race in 1973.

The best examples of use of the various materials are as follows:

Steel Large motor-yachts (i.e. small ships) and strong heavy vessels such as Moitessier's *Joshua*.
Light alloy Racing yachts where cost is less of an object than rigidity and high performance, high-speed power-boats (12-metre *Courageous*).
Wood planking Individual vessels of aesthetic appeal. Also in vast numbers existing craft, built before 1960s (examples everywhere).
Moulded wood One-off racing boats, sail and power, and cruising boats of any type when the owner is prepared to pay for it.
GRP All types of modern series produced boats and yachts from the smallest to about 100 ft (examples everywhere).
Glass-fibre foam sandwich One-off system which enables rigid boat with advantages of plastic construction to be made without many moulds (*Great Britain II* and *III*).

RIGS AND EQUIPMENT

The most efficient use of the wind on ships in the history of sail is found in modern yachts rigs. Yachts have fore-and-aft rigs and from antiquity this has been one of the two basic systems of sailing on water, the other being the square rig. Almost any rig will blow downwind, but only recently have seagoing vessels (ocean racers) been able to sail as close to the true wind as 35 to 40 degrees.

The angles to the wind which certain types of sailing-vessel could point are:

Seventeenth-century warship	90 degrees
HMS *Victory*, early nineteenth century	70 degrees

Clipper ship, late nineteenth century	65 degrees
Pilot cutter, late nineteenth century	55 degrees
Cruising yacht, early twentieth century	45 degrees
Ocean-racing yacht, 1975	38 degrees

The first three ships above were all square rigged, the remainder fore and aft.

The first fore-and-aft sail in the world was probably that used by the ancient Egyptians on the Nile 'nagger' in the fourth century BC. In shape it is nearly a squaresail (the rig can still be seen on the Nile today), but it has been tilted so that one bottom corner forms the tack of the sail. Thus the sail can pivot as the boat tacks. It is this manœuvring ability to tack through the eye of the wind that distinguishes the fore-and-aft sail from square rig, rather than close-windedness. The latter is a much later development. The nagger at 85 degrees to the wind is little better than if she was square rigged.

Fore-and-aft rig as used in working boats in the 19th century. It is typical here in pilot cutters who raced to be first to put a pilot aboard a ship: it was hardly surprising that early yachts racing used the fore-and-aft rig.

The main emphasis on commercial and naval shipping from Europe through the centuries has been on square rig, because these are the ships that were able to take the routes of the trade winds and prevailing winds. The fore-and-aft rig was not suitable for running long distances in the ocean. (Some yachtsmen thinking of spinnakers would agree that this is still true today.) However the Arab dhow with its lateen sail has used the fore-and-aft rig for centuries, as has the Chinese junk. In northern Europe fore-and-aft rig was used in the form of a sprit-sail (which today survives in the Thames barge) and later the gaff mainsail, as finally seen on sailing pilot cutters at the beginning of the twentieth century (for example *Jolie Brise*).

The most ancient sail seen on modern yachts is the jib – a triangular headsail. In all the developed square rigs and ancient ones, the jib, a fore-and-aft sail, was set between and forward of masts carrying squaresails. There is now no distinction between jib and staysail except in their respective positions (if there are two headsails the outer is called a 'jib'), but jibs developed in the fifteenth century as rigs developed and multiple headsails were set forward on bowsprits, which had originally been constructed to carry little squaresails under them ahead of the bows (watersails). The word 'jib' is a derivation of 'gibbet', which implied a sail hung from part of the mast.

Spars

The universal material for masts and spars was wood until the nineteenth century. One has only to see a straight tree to realize that here was the ready-made 'stick'. The first patent for an iron mast was taken out in 1809 by Richard Trevithick and Robert Dickinson of London. In 1825, the first naval experiment was made, the 46-gun frigate *Phaeton* being fitted with an iron mainmast and iron bowsprit. The mast was of hollow wrought-iron cylinders, 3 ft 6 in long, riveted together. In the 1860s steel masts came into general use on commercial sailing-ships.

The first metal mast on a racing yacht was the steel one designed by Herreshoff on the America's Cup defender *Columbia* in 1899. For yachts generally, wood was lighter, especially as the art of spar-making involved hollowing out wood (spruce was favoured) and then fastening approximately semi-circular sections together to give a hollow wood mast with the essential lightness required.

The first light alloy mast used was on the 1937 America's Cup defender *Ranger*. The Bath Iron Works, of Maine, built to the design of Olin J. Stephens a 'duralumin' spar of 165 ft. It was supported by rod rigging in panels of 17 ft maximum length as this was the limitation of the drawn rod. When on tow from the builders to Newport, R.I., the rigging unthreaded aloft and the yacht was dismasted. The expensive mast sank and had to be replaced.

The first extruded metal mast came in 1948 after numerous light alloy masts had been made by welding or riveting formed plates. The extrusion was designed by yacht designer John Powell (Brit.) and colleagues of the Sussex Shipbuilding Company, Shoreham. The extruding was done by Birmabright of Birmingham, who had been supplying light alloy for marine purposes since 1929, such as metal dinghies, and Second World War patrol boats. The mast was 65 ft from deck to truck and in three extruded sections arc-welded. It was fitted to the 55 ft ocean racing yacht *Gulvain*.

Satanita, which had a main boom of 92 ft and mainsail area of 4919 ft^2.

The longest main boom ever fitted to a British yacht was the 92 ft spar of Oregon pine on the mainsail of the 131 ft 6 in cutter *Satanita* built in 1893 for A. D. Clarke (Brit.) by J. C. Fay of Southampton (now a yard of Camper and Nicholsons), to the design of J. M. Soper. The mainsail was 4919 ft² and displacement was 130 tons. It was *Satanita* that sank the cutter *Valkyrie II* on 5 July 1894, when racing on the Clyde. Even if there was such a phenomenon as a single-masted yacht of such a size today, the main boom would never be as long because of the shape of modern rigs (the mainyard of the clipper ship *Donald Mackay* was already exceptional at 115 ft).

The spars of new yachts today with very few exceptions are of extruded light alloy in standard sections. Sections are used for masts, booms, spinnaker booms, auxiliary spares of various sorts, and even short flag masts for motor cruisers. One or two isolated yards may have craftsmen who can still build a wooden spar, but the art is virtually dead. The modern alloy spars come from specialized manufacturers and masts are tapered and anodized silver, gold or black. A leading spar-maker in 1975 had stock extrusions for 15 yacht masts, 8 centreboard dinghy masts, 7 yacht booms, 4 dinghy booms, and about a total of 40 extruded sections which included tube spinnaker poles, spreaders, and auxiliary sections.

The tallest yacht mast in light alloy manufactured since 1945 was a 136 ft 6 in spar for the Greek steel schooner *Vagrant*, 109 ft, built by Proctor Masts of Southampton, England. The cross-section of the mast was 21½ in × 14 in and a man was able to crawl through it to adjust halyards and wiring. Although extruded, the extrusions were quarter sections which were then riveted together. *Vagrant* was of 1913 build and designed by Herreshoff.

The longest spinnaker boom in light alloy to be built in Britain was a 42 ft pole for a spinnaker of 10 200 ft² on another Greek ketch, *Kettyioana*.

The most favoured cross-section for yacht masts is the ellipse. The requirement is to minimize interference from the mast with the airflow over the mainsail, yet the mast must still be massive enough to hold up the rig. There are alternatives in degree between a big section with thin walls and a small section with thick walls, but if the latter is carried to extremes it becomes too flexible and heavy. Various shapes are tried experimentally. Thin strips on the sides of masts, for instance, have been used to improve airflow.

The most advanced masts are used on C class catamarans. They merge into the fully battened single wingsail and, being in effect 40 per cent of the sail area, turn with it, thus cutting out the bulk and windage of a normal mast. Such masts are banned by most other classes in their rules to save cost and ensure seaworthiness. Revolving masts are banned in racing rules unless specifically permitted in class rules.

Sails

The universal materials from which sails are made on modern yachts are two groups of man-made fibres – Terylene and nylon. Spinnakers are made from nylon (which has appreciable stretch) and fore-and-aft sails from Terylene. The commercial name for Terylene in the USA is Dacron; in France, Tergal; in Japan, Tetron. Yacht sails come from specialist firms and there is considerable variety in quality of cloth, cut, and systems of panelling, shape, and price. Some firms actually weave their own sail-cloth in Dacron (notably Hood, USA, and Ratsey and Lapthorn, England).

The earliest sails in antiquity were woven from rushes, reeds, and grass and then animal skins were used: rawhide still is used today to reinforce corners and for selected patching! Hemp for sails was introduced by the ancient Greeks, but flax was the material used for a thousand years until the nineteenth century. Flax was baggy and the shape erratic, so when *America* arrived in Britain in 1851 with flat sails of cotton, it caused a sensation and changed the fashion. Cotton sails were then used on yachts for just about a hundred years. They were subject to rot and had to be carefully stretched when new.

The oldest yacht sailmaker still in business – and actively so – is Ratsey and Lapthorn of Cowes. A separate firm under members of the Ratsey family who live in America, has also long been established at City Island, New York. Lapthorn was making sails in Cowes at the end of the eighteenth century and HMS *Victory* carried a Lapthorn fore topsail in 1805 at the Battle of Trafalgar, but previous generations of Cowes men of the same name had been making sails of flax.

The first sail of any man-made fibre was the genoa on board *Ranger* in 1937. The material used was rayon. Nylon first appeared in 1939 in the USA and was tried out for sails in 1945. It was soon found to have too much stretch to enable jibs and mainsails to hold their shape in a breeze: it has been used for spinnakers where requirements are different, with success ever since.

The first polyester fibre, from which Dacron is made, was produced commercially in the United States in 1953. It was first known as 'Fibre V' and later as 'Dacron Polyester'. Polyester fibre had been invented in 1941 as polyethylene terephthalate by Whinfield and Dixon of the Calico Printers Association, Manchester, England.

The first Terylene sail to be made in England was a genoa for the 8-metre CR class *Sonda* by Gowen of West Mersea in 1952.

The first big success of Dacron was in the 1954 Star World Championships in Portugal. The Cuban helmsman Cardenas, who had the only sails in the fleet of Dacron, won a clear 1st place. The first yacht to carry a complete inventory of Dacron sails for long-distance racing was Richard Nye's *Carina*, 53 ft, in 1955 in which year the yacht won the Fastnet race. Specially woven and treated Terylene as used in sails today has immeasurable advantages over earlier materials. It can be used immediately without losing its shape; cannot rot, though mildew does appear on it in certain circumstances without rotting it; it can be adjusted for shape with suitable control lines; it is soft to handle, fold up, and stow; it is very long lasting; storm sails, which were formerly very thick, can be more than adequately strong of easily handled, comparatively light material; it can be made in sizes down to the very light weight for light air sails and nylon can be as light as $\frac{1}{2}$ oz/yard^2. The only disadvantages found with synthetic materials after 25 years development are that, firstly, they will deteriorate in strong sunlight, due to ultra-violet rays, and, secondly, that the stitching stands proud of the hard cloth and is, therefore, subject to chafe. The first problem can be overcome by keeping the sail covered or stowed away when not in use so that direct sunlight does not beat down on it – for instance, when the mainsail is folded along a boom; and the second problem can be alleviated by rows of extra stitching and by making sure that any weak stitching is immediately repaired.

The most popular rig on new boats is the masthead sloop. This applies to all small cruisers, ocean racers, and day boats. Many smaller one-design classes and centreboard dinghies are also sloop rigged, but the rig only going a part of the way to the masthead. Other rigs are often found in older boats and larger yachts. For larger yachts a ketch rig breaks up the sail area into manageable proportions. Once a yacht is over LOA 50 ft, the sizes of the individual sails in a sloop become very large and so a mizzenmast means that the mainmast no longer 'grows'. Mainsails and mizzen-sails are invariably Bermuda rigged.

The actual rig of Bermuda in the 19th century, which later became most commonly used rig of all, the Bermuda rig. The exciting boats with tall triangular rigs beat to windward with native crews. The mainsail was laced to the mast and could not be lowered or reefed under way.

The greatest step towards the simplicity of the rig is the Bermuda mainsail. Until 1920, the rig was scarely seen, and yachts were gaff rigged. This involved, or involves, since there are many still about, having an additional halyard to pull the gaff up to the top, a large spar which is then above the head of the crew until such time as the sail is lowered. When extra canvas is desired it is necessary to hoist a topsail above it, which again means more 'strings'. The short mast also means that headsails cannot be as long in the luff as is desirable for good performance, especially in good weather. The main boom is also liable to be out beyond the stern, where reefing and other operations are dangerous. In its simplest form the Bermuda sail only requires one halyard to pull a triangular sail up a groove or track and the boom is well inside the yacht because sail area can be spread upwards, owing to the useful height of the mast. The period in which yachts changed over to Bermuda rig can be judged from the fact that the 1920 America's Cup was sailed with gaff rigs and topsails, while the next contest in 1930 was sailed with Bermuda mainsails. The first Fastnet fleet in 1925 was all gaff rigged, some of the boats having squaresails for running.

The first attempt to preserve the gaff rig is now undertaken by societies interested in vintage yachts. Among those are the Old Gaffers Club in England and the equivalent associations elsewhere. The practicality of the modern Bermuda rig does mean that there is a certain sameness about sailing boats seen at other than very close. Preservation of gaff-rigged boats gives a great pleasure to individuals and is also a welcome variety in yachting waters.

The first Bermuda rig was derived from the triangular sail on a spar called 'a shoulder-of-mutton sail'. It was used as a small mizzen on ships' boats in the British Navy and its luff was laced to a single mast. This loose-footed sail was sheeted to the transom.

The first Bermuda sail appeared in 1808 in the Islands of Bermuda when a local resident called Harvey rigged his local schooner with very high lateen-like sails instead of the usual gaff rig. The masts were heavily raked. The schooners of Bermuda of the early nineteenth century were unique in the sailing world in having not only the mainmast, but all masts, with triangular loose-footed sails. Of course, there were no tracks or elaborate crosstrees and the sails were laced or had hoops in the manner of gaff rig to hoist them.

In the following years the Bermuda rig developed on sloops. A contemporary description said 'All Bermudian races consist as much as possible in beating to windward and on the morning of the regatta or match, two marks are laid down in Great Sound. One of these is dead to windward and three miles distant. The boats sail from leeward round it and win at the weather mark on the second round. Off the wind the yachts are slow and in strong breezes run badly, sometimes burying themselves and always acting so queerly that Bermudians will not gybe them. The mainsail is laced taut to the mast and cannot be lowered and as it can be neither doused

The first Bermuda rig on a large yacht was fitted to *Nyria* in 1921. Note multiplicity of small jibs (sagging badly), massive rigging and long boom when compared to pictures of modern yachts.

Right
Solent One Design Class early in the century with different rigs. Some yachts have almost achieved the Bermuda rig with a high-peaked yard while others still have gaffs and topsails.

or reefed, the boat must carry whole sail unless she is swamped or the stick goes overboard.'

Another observer said 'a 50 ft mast in a 25 ft boat with a 34 or 40 ft boom was not uncommon. A coloured skipper was employed who would enroll his friends, dress them in striped jerseys and caps, and put them up to windward, over a ton and a half of shifting ballast, serve out a lot of rum all round, off they go generally with the head of the mainsail lashed to the mast-head so that she must carry her whole sail all through the race or swamp.'

There was a smaller dinghy class also using Bermuda rig at Hamilton in 1883, where the boats were LOA 14 ft, beam 4 ft 6 in, with a 25 or 30 ft mast and a 25 ft boom and 15 ft bowsprit. The Royal Hamilton Dinghy Club is still active in Bermuda. In this class 'bathing costume is considered the correct thing and is well adapted to the climate; it is also desirable in this sport to be able to swim as there is no room in the boats for such superfluities as lifebelts'.

The first racing yacht in America to be designed with Bermuda rig, that would give specific advantage, was the half-rater *Ethelwynn*, designed by W. P. Stephens in 1895. She successfully defended the Seawanhaka Cup against her gaff-rigged opponents in 1896. Yachting writers of the day disapproved of the rig and defended the gaff, while leading yacht-designers ignored it and continued to try and perfect the gaff rig.

It was still possible to have arguments about the merit of gaff rig for cruising yachts and ocean-cruising yachts as late as 1950. But today gaff rigs are only sought as are vintage cars. In 1917 a new one-design, the Larchmont O Class, 45 ft, came out with gaff rig and topsail, though one of the class had an experimental Bermuda mainsail. The popular Star class began in 1908 with a gunter rig and did not change to Bermuda rig until 1922. (The rig has since been modernized on two occasions.)

The first large racing yacht to carry the Bermuda rig was *Nyria* in 1921. She was a British boat of the 23-metre class to the International Rule, owned by Mrs R. E. Workman, and the rig was designed by Charles E. Nicholson.

The early Bermuda rigs, such as on *Nyria*, were sometimes called 'marconi' because the masts were a mass of struts and wires like a radio transmission mast. This term was used more in the USA than Britain and in the USA the terms 'jib-headed mainsail' and 'leg-of-mutton sail' were also used.

Genoas

Apart from the Bermuda mainsail, the other sail which gives the modern rig its main characteristic is the overlapping genoa jib. Some dinghy and one-design classes do not allow this, but it is present in some of the leading ones, such as the International 14 and the Flying Dutchman. The Soling class, for instance, has a one-design jib which hardly overlaps the mast. International Offshore Rule boats and cruising boats carry genoas. For cruising the genoa is a very useful light-weather sail, for its large area is all set inboard: this is in contrast to the ancient method of setting light-weather canvas whereby jibs were set to the end of bowsprits or topsails hoisted to the masthead.

This was indeed the origin of the sail because working boats like Brixham trawlers were able to carry 'reaching headsails' in light or moderate winds with the wind abeam. The exact shape was not too critical, but the extra area was especially useful.

The first genoa jib (known today as a 'genoa') evolved from these early light reaching sails. The sail was first seen in the International 6-metre class at Genoa Regatta in February 1927. The Swedish helmsman Sven Salen's boat *Lilian*, in common with several other 6-metre boats, had a 'balloon staysail' for extra area when reaching. However, on arriving at the leeward mark Salen hardened in what would now be called a genoa and this is the origin of the term. Later in 1927 Salen sailed the 6-metre *Maybe* at Oyster Bay in Long Island Sound.

This conversion of the balloon staysail for light weather reaching into a flat sail for moderate and in later years hard weather beating, came to be widely adopted. In recent years, the genoa has in effect become the mainsail, because Terylene has enabled even larger genoas to be carried and hold their shape in tough conditions.

Spinnakers

A sail with the greatest influence on the handling and in the design of the yacht for performance off the wind is the spinnaker. The basis of the modern spinnaker is that it is a symmetrical sail of something between a triangle and a sphere in shape, the proportions and dimensions relative to the rest of the rig being limited by the rules of the yacht's class. Because its area is larger than any other single sail it has been exploited in recent years so that it can be used not only when the yacht is running, but when reaching and even with the wind forward of the beam. This is achieved by various cuts, which enable it to drive the yacht forward on these points of sailing despite its 'bagginess'. Various sailmakers have trade-names for different particular spinnakers that are developed from time to time, such as 'star cut' and 'tri-radial'.

The first spinnaker was found to be effective when it was tried on a yacht instead of a squaresail, the traditional sail for running. The first spinnaker was hoisted on 5 June 1865 by the yacht *Niobe*, owned by William Gordon, in a match race off Cowes. *Niobe* hoisted a large triangular sail to the topmast head and boomed it out, which resulted in her rapidly drawing away from her rival. Gordon was a sailmaker of Southampton and in 1866 sold one of these sails to Herbert Maudslay, owner of the yacht *Sphinx*, 47 tons. *Sphinx* used the spinnaker at the Royal Yacht Squadron Regatta of that year and the term 'spinnaker' started to appear in the yachting press from then on. The term was not known in commercial vessels or fishing vessels, so is widely thought to have originated from the name of the yacht *Sphinx*. However, the grandson of the skipper of *Niobe* wrote in his memoirs that his grandfather said that a hand on board the yacht commented 'that is the sail to make her spin', when it was hoisted. One of the leisured afterguard apparently coined the phrase 'spin-maker', but in any case the term was soon in general usage, in the same way that new developments in other gear and sails quickly receive names and are widely used in yachting today.

The most significant step in spinnaker development occurred, like the genoa, in the International 6-metre class. The early spinnakers were triangular and set completely to windward of the forestay. It was, therefore, only possible to carry them when running nearly dead before the wind. In 1927 on *Maybe*, Sven Salem set a symmetrical spinnaker of which the sheet was taken to leeward of the forestay and then under the main boom. This is the original of the spinnaker as known today and was originally known as a 'parachute spinnaker' or 'double spinnaker'. It is easy to gybe because whatever arrangement is used the sail remains symmetrical although the mainsail comes over when gybed. Ocean racing has bred the 'masthead spinnaker', which is the same sail set from the masthead to give greater area and a wide variety of various ways of arranging the seams.

The big boy, or blooper, a development carrying in effect a second spinnaker; the practice originated in New Zealand in 1971.

The first appearance of second spinnaker-type sails carried simultaneously with a normal spinnaker was in New Zealand in the 1971 One Ton Cup. The New Zealand yacht *Wai-Aniwa*, skippered by Chris Bouzaid, a sailmaker, hoisted a full genoa to windward of her spinnaker. This was the logical result of the development of the spinnaker from a sail set to windward to one set symmetrically and then finally to leeward. The gap to windward was then plugged by this further sail which has become known as the 'blooper' (USA) or 'big boy' (Brit.). Special bloopers are now developed, though they have to accord with the rules which limit genoas and jibs generally.

Largest sails

The largest rig ever placed on a single-masted vessel was the 16 160 ft² of *Reliance* in 1903 (see Section 3).

The biggest area of mainsail ever made was also on *Reliance*. It was 6334 ft² and required 1¼ tons of Egyptian cotton to make it.

The largest sail made in Britain since 1945, *Creole*'s spinnaker seen on the floor of its sail loft.

The largest sail ever made was the spinnaker of the J-class America's Cup defender *Ranger* (see Section 3) in 1937. It was 18 000 ft² (or two-fifths of an acre). Her genoa was also the largest at 4320 ft²: when the wind rose to more than 12 knots this headsail was replaced by two smaller headsails, which maintained their shape and could be handled by the crew.

The largest sail made in Britain since 1945 was a spinnaker for the schooner *Creole* owned by Stravros Niarchos in 1957. The sail was made by Ratsey and Lapthorn of Cowes and had a luff of 122 ft 8 in (being symmetrical, the leech was the same) and the width across the foot was 90 ft. The total area was 8555 ft² and the material used was 5 oz nylon sailcloth. The sail was too large to lay out on the loft floor in the usual way, so had to be laid out in sections and finally sewn together. It was sewn throughout by machines, though the corner patches were sewn by hand for extra strength.

Modern gear

Increasing numbers of auxiliary sails and sail-adjusting gear have swelled nautical vocabulary as a result of competitive sailing. In old and not so old nautical dictionaries will be found words like 'bowsprit', 'bulwarks', and 'poop', but these parts of the vessel have no place in a modern sailing-boat. Examples of modern language referring to rigging and sails are:

Barber hauler Line which moves the genoa sheet block athwartships.
Bendy Boom Main boom of suitable section so that when a centre mainsheet pulls out it curves and flattens the mainsail, which is needed in fresh winds.

Cunningham hole Cringle in the luff a short distance above the tack of a sail. Hauling down hard on it with winch or tackle stretches the luff and flattens the sail. Mast bend also flattens the sail and a number of classes have masts designed to be bent as required.
Foot zipper When the mainsail is flattened the loose cloth at the bottom is gathered up by using a heavy-duty zip-fastener.
Ghoster Very light-weight headsail for use in light airs (say up to 5 knots of wind).
Kicking strap – 'boom vang' in USA A tackle, wire or solid, adjustable strut which pivots from the mast to a point on the boom to stop it lifting and spoiling the shape of the sail. It originated in dinghies and spread to all sizes of boat.
Leach line Thin line sliding in a special seam in the leach of sails, adjusts shape and cuts out shake along the edge.
Single-pole gybe Implies gybing the spinnaker by swinging one pole across the foredeck, leaving the old tack which becomes the clew and attaching to the new tack. There are many techniques for gybing the spinnaker including the 'double-pole gybe'.
Slab reef The modern name for 'ordinary reefing' – reduction of mainsail size in freshening winds. Sail is secured by cringles higher than the usual tack and clew thus removing an area of sail. *Jiffy reefing* is a quick variation; *through the mast reefing* is rolling the sail round the boom by a handle whose axis passes into the boom end, through the mast; old pilot cutters and many later yachts had roller reefing with worm gears between boom and mast which made rolling down the mainsail difficult and produced an unwanted shape.
Spinnaker chute The quickest means of hoisting and lowering a spinnaker is by pulling it out of a funnel which leads into and along the deck. The spinnaker is thus ready for instant use. This technique is only possible so far on boats under 25 ft including keel boats and centreboarders.
Stretch luff genoa Genoa with heavily taped luff of Dacron which responds well to luff tension applied by halyard winch. Hardening the halyard flattens the sail. The result is that the sail can be used effectively for wider range of wind speeds than the old-type genoa with a wire luff rope sewn in.
Turtle Typical among newly invented words of which the meaning cannot be apparent. It is a canvas or plastic holder for the spinnaker, secured on deck and out of which the spinnaker is hoisted. The spinnaker has to be repacked in it later in the cabin (in a large boat) or on shore (in a day boat). Alternative methods of hoisting are the chute (see above) and *Stops* in which the spinnaker is secured by elastic bands in a long sausage and then hoisted and 'broken out'.

Spinnakers are frequently put into 'stops', so they can be hoisted like a sausage and the wind will not get into them until they are broken out. Here a spinnaker is stopped by passing it through a bottomless bucket and applying elastic bands.

The constructional feature that marked the parting of the ways between sailing yachts and sailing commercial vessels was the outside ballast keel. In commercial vessels cargo was stowed to give stability and in yachts which were not burdened with cargo, stores or gravel were carried low to give power under sail. It was the tonnage rating rules that forced the designers to try and sling the ballast even lower on the narrow yachts of the 1870s and 1880s, (see p. 190). The designer George Watson wrote at the beginning of this century: 'About this date 1875 – builders were becoming more and more impressed with the value of a low centre of gravity got by outside lead, which in combination with increased displacement allowed of beam being reduced and length added almost indefinitely . . . the older men were naturally timid about the introduction of external ballast and it was left to "the boys" with the happy audacity and confidence of youth to design 100 tonners with 70 ton keels which perhaps fortunately did not get beyond the length of paper. But fives, tens and twenties [tonners] were built with nearly all their lead outside and did not from that fact tear themselves asunder. James Reid of Port Glasgow, designer . . . of many fast boats and later the writer [Watson] put all, or nearly all, of the ballast outside and the practice in a few years became general.'

The first 'application of outside ballast' was reported by George Watson as being in 1834 in a yacht called *Wave* owned by John Cross Buchanan of Glasgow. 'There may have been earlier instances of this in the south [of England]', but *Wave* had a metal keel fixed to her.

Inside ballast is by contrast today considered positively dangerous, because if the yacht is knocked down, it could break loose and cause appalling damage. Some inside ballast, in the form of lead pigs, is sometimes carried, well secured for trimming purposes. The modern lead keel is moulded densely and the best designers pay considerable attention to the hydrodynamic shapes of the section (in plan) and the extremities, like the wing-tip of an aircraft.

The most dense ballast keel was that fitted to *Pen Duick VI* (Eric Tabarly) before the 1973 Round the World race, being lead containing inert uranium. After that materials denser than lead were banned under the IOR. Lead is the most common ballast, but cast iron is used widely to save cost. The point is that the greater the density of the material, the less its volume, the less its wetted surface and the lower the centre of gravity in it can be.

Not least among the appeal of sailing to the modern sailing city-dweller is the feeling of pulling on ropes in the open air rather than pushing buttons behind glass and concrete. Ropes continue to be the primary method of controlling the rig of a sailing yacht, though their complication has been reduced from the vast amount of cordage seen on old sailing ships. Mechanical means in the form of winches are, however, used to handle sheets, halyards, and other lines quickly and conveniently. For hundreds of years sailing ships and small vessels used *manpower* as the chief method of hauling a line: enough members of the crew would 'tail' to hoist a sail and so on. This was aided by *tackles*. Anchors and mooring warps were moved with the aid of capstans. A capstan was a rotating drum with bars ranged round it so that once again as many seamen as possible could walk round it.

The winch on the other hand was a smaller mechanical device that made use of the leverage handle and the friction of the drum. One more method of tightening a rope, often used in smaller craft, is the jig ratchet where after a rope has been tightened as far as possible, it is pulled sideways across a ratchet to tauten it.

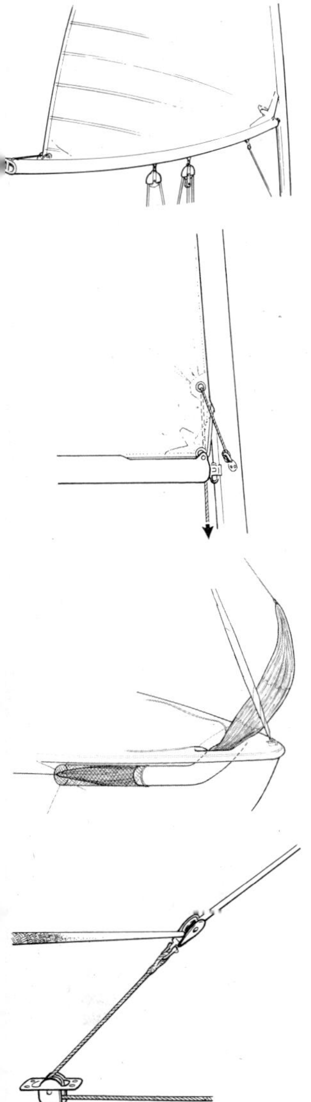

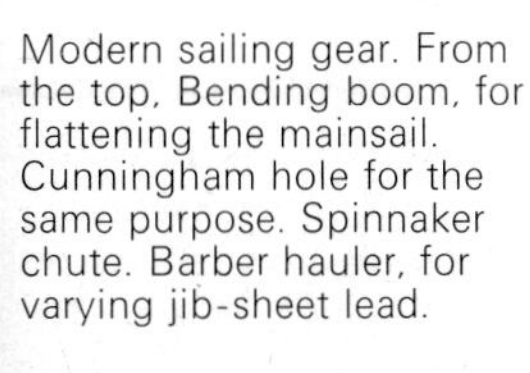

Modern sailing gear. From the top, Bending boom, for flattening the mainsail. Cunningham hole for the same purpose. Spinnaker chute. Barber hauler, for varying jib-sheet lead.

Winches

The most useful device to survive from previous methods of handling under sail is the *winch* (a sixteenth-century word). Tackles are still used, but since 1950 their use has been reduced on new boats. The disadvantage of tackles was the weight of blocks and purchases aloft, the great lengths of line required to obtain a mechanical advantage, and some danger from moving blocks such as on the clews of jibs where a pair of blocks would thrash about as the yacht was tacking.

The most common type of standard winch up to 1950 was the bottom action one made of bronze available in sizes up to about 5 in diameter. This was a direct action winch where the mechancial advantage came from the leverage of the handle in one direction and the ratchet holding it in the other. A number of manufacturers made these. As they were not powerful they were often used in conjunction with tackles.

The most sophisticated winches up to that time were custom designed and made; pedestals and 'coffee-grinders' had been used in the J-class yachts and *Ranger* had one with two gears for hauling in genoas at different speeds.

The simplest type of winch is the snubbing winch. This has no handle, rope being pulled round it and a ratchet and the friction of the winch holding the rope on which tension must be kept by the crew. This is used in very small boats and such winches are often made of laminated plastic. After 1945 laminated plastic was tried for larger winches and marketed, but, in anything but the smallest sizes of winch, it does not stand up to continued use.

The first yacht to put sheet winches below deck, in an effort to get the weight lower, was the American 12-metre *Valiant* in 1958. At the same time one of her competitors *Weatherly* was the first yacht to have more than two coffee-grinders on board.

Modern cross-linked winches with variable gears, which can be driven by two men winding the centre pedestal. Switch (seen below pedestal) enables winding to be applied to one winch or the other. On left are smaller winches for various purposes including halyards, forward guys and other gear: the main winches are reserved for sheeting the genoa.

The first cross-linked winch was fitted on the Australian 12-metre *Gretel* in 1962. It was designed by Alan Payne (Austral.). A cross-linked winch is one whereby handles on one side of the boat will turn the drum on the other side of the boat. This has been elaborated in various forms to enable more than the usual pair of men who work a pedestal winch to wind the drum, two pedestals exerting their power. In smaller versions entire winch systems have been linked round cockpits and this was first developed to full extent in *Frigate*, owned by Robin Aisher in 1973. With a total of four winches and six handles, any one or more handles could drive any winch, thus exerting as much power as required and distributing the crew around the cockpit for the best technique in sailing the boat.

The major stages in the development of yacht winches have been, in this order, the use of wide diameter drums linked to gearing to combine friction and power; the operation of two gears by winding the handle in opposite directions, thus abolishing small control levers; adoption of the

Spinnaker halyard broken and the sail blown away: hardly to be rated as an accident, but more as a routine mishap when racing

top action sheet winch so that a man can wind a circular motion rather than a slow ratchet manner; and linked action as mentioned above.

The greatest encouragement to the use of winches was the introduction of Dacron lines, thus enabling easily handled material to be used on these winches. Previously, natural fibres had kinked badly when handling and were unable to stand the big strains imposed by winches. To put it in another way, the great strains that small diameter synthetic rope can stand today have enabled more powerful winches to be developed. Because of the low breaking strain of natural fibre, wire was used in larger boats and this was dangerous and difficult to handle.

The most favoured materials for winches are now stainless steel, bronze, and sometimes titanium. Light alloy is used to save weight but requires more maintenance. The efficient winch is one of the reasons for the development of big genoas in IOR boats and cruisers. A 45 ft yacht with two sheet and two halyard winches would not have been uncommon in 1950: now such a yacht, even if a standard production one, would not seem strange with 20 winches of many shapes and sizes.

Most of the devices seen today as standard were originally made as prototypes in top racing boats. Designers used to draw up mechanical specifications and designs in their drawing-offices and have them put together. Today the manufacture of winches is a specialized technique for firms with their own drawing-offices and marketing departments. Such firms will also make prototypes for top racing skippers as part of their research and development. Such firms include Barient of San Francisco, Lewmar of England, Barlow of Australia, and Goiot of France.

The largest standard direct action winch is the Barient No. 35. Made of stainless steel it weighs 65 lb; of aluminium 40 lb. It has height 11 in, drum diameter 5¾ in, and a lowest gear ratio of 54·5 to 1.

The most expensive single-winch system installed in Europe is the twin coffee-grinder and twin pedestal-linked winches made by Barient and installed in 1975 on *Gitana VI*, owned by Baron Edmund de Rothschild. The base of the coffee-grinder drums is 1 ft 5 in and height 1 ft 1 in. The height of each pedestal is 2 ft 5 in from the base of the equipment. The cost of the equipment, which was £6000, was mainly due to the elaborate three-speed-gear equipment.

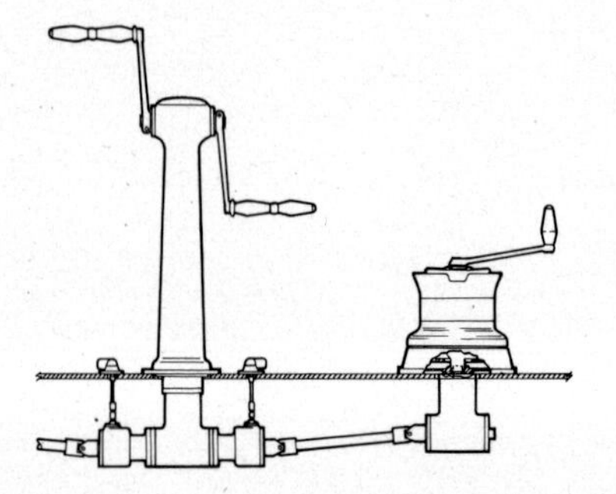

A standard sheet winch system of 1975. The double-handled pedestal enables two men to sheet in the genoa very fast. The handle on the winch itself would be used for trimming. The second winch would be sited to the left.

The most compact pedestal and coffee-grinder to be made as standard is by Lewmar and called the 'mini-grinder'. Pedestal and grinder come in a single-base casting; there is a three-speed-gear change and the best power ratio is 50 to 1. The grinder will take a load of 10 000 lb.

6 Grandeur and Grief

ROYAL YACHTS

The Royal Yacht *Britannia* at her mooring in Cowes Week. Passing her is *Coweslip*, the Flying Fifteen class boat owned by Prince Philip.

The largest royal yacht in the world and also larger than any previous British royal yachts is HMS *Britannia*. She is the world's largest yacht today, but this is not comparable even with the large private motor yachts (Section 5) because she is a naval vessel in the Royal Navy and is intended to convert into a hospital ship in time of war. She has been much criticized for the cost of upkeep and refitting since she was launched by Queen Elizabeth II on 16 April 1953. In charge of the royal yacht is an Admiral and the crew by tradition wear soft-soled shoes and no orders are shouted. She is one of the world's most impeccable ships.

The nearest which *Britannia* comes regularly to other sorts of yachting is during Cowes Week when members of the Royal Family are on board. Queen Elizabeth has, however, never been present at Cowes in any yachting connection. *Britannia* was specifically named after the royal racing cutter which was scuttled in 1936.

The length overall of *Britannia* is 412 ft 3 in, beam 55 ft, displacement 4715 tons. She was built in steel by John Brown and Co. of Clydebank and has four steam turbines of their construction, giving a total of 12 000 hp and maximum speed of 23 knots. The vessel was completed in January 1954 and first conveyed the sovereign from Tobruk via Gibraltar to London in May 1954. The yacht was first planned in 1939 but shelved until after the Second World War. Decoration in the yacht is generally restrained but there are some relics from former royal steam yachts to add embellishment.

The most distinguished of royal sailing yachts in history was the racing cutter *Britannia*, 122 ft. The introduction of steam meant that sail subsequently was only used for conveying the royal family for sporting purposes. The yacht was built for the Prince of Wales (later Edward VII) to the

King George V at the wheel of *Britannia*.

H.M. YACHT VICTORIA & ALBERT.

'Britannia'

1893 to Aug 4th 1924

Starts	1st Prize	Other Prizes	Total
323	164	44	208

King George V wrote down *Britannia*'s record in his own handwriting.

design of George Watson of Glasgow. Built to the length and sail area rule, she rated at 151, but was never intended to be a rule-cheater. The result was that she was measured over the years to several different rules ending up in the 1930s as 76 ft rating under the American Universal Rule (J class). Her other dimensions were LWL 87 ft, beam 23 ft 3 in, draught 15 ft. The sail area varied each time the rig was altered: in 1893 it was 10 797 ft^2 and in 1933 it was 9235 ft^2. *Britannia* was not in royal ownership throughout her life and was sold in 1897, then repurchased in 1899, sold again in 1900 and bought by King Edward VII in 1902. After his death King George V made active use of her.

Britannia won a record number of prizes in her time. This was not so difficult as in first-class racing today, partly because the entries were small and partly because *Britannia*'s racing life spanned 43 years. On 4 August 1924, a yachtsman at Cowes was able to ask the King how many prizes the yacht had won. King George V wrote down on a sheet, the facsimile of which is preserved in the RYS Library, 'Britannia, 1893 to August 4th 1924 Starts 323, 1st Prize 164, other prizes 44, total 208.' Her final score was the greatest of any yacht in racing history: 625 starts, and 231 first prizes.

By order of King George V and because none of his sons was interested in the sport of yachting, *Britannia* was scuttled after his death. Every moveable fitting was taken off and auctioned for charity. Many yacht clubs now have one piece or another on show on a wall or in a glass case. Uffa Fox and very few people including a small group of schoolboys helped launch the stripped hull from Marvins Yard at Cowes in July 1936: then the yacht was taken by two destroyers from Portsmouth and sunk south of St Catherine's Point, Isle of Wight.

The first royal yacht was also the first yacht in England: the 50 ft *Mary* (beam 18 ft 6 in) owned by King Charles II in 1660. Her origin has been recounted in Section 2. It should be added that she was used for match races as were other of King Charles's yachts. After 1685 royal yachts were only used for conveying the sovereign, for naval use, and for some relaxation in the sense of taking the King for a sail – but not again for racing until *Britannia*.

The only sailing yacht owned by British royalty today is the Flying Fifteen *Coweslip*. This 20 ft keel boat was presented to Prince Philip by the town of Cowes in 1948. The yacht has been frequently raced and kept at the slipway of Uffa Fox, until the latter's death in 1972. Queen Elizabeth and Prince Philip have also owned *Bluebottle*, 29 ft 3 in, a Dragon class boat presented as a wedding present by the Island Sailing Club, Cowes, in a successful effort by leading yachtsmen to lure royalty back to Cowes, where it had not been since the death of King George V. From 1947 to 1965, *Bluebottle* had a Naval Sailing Master appointed and raced regularly at home and abroad in Dragon regattas. The second sailing-yacht to be owned by the Queen and Prince Philip was the ocean racer *Bloodhound*, 63 ft 5 in. This pre-war boat was extensively refitted in 1961 (see p. 204). She was used by Prince Philip for racing in Cowes Week and for some cruising, but when not being sailed by him she was on a programme of sailing and racing for club members and young people. She was sold in 1970. Since then Prince Philip has been loaned the use of an ocean-racing yacht for several days in each Cowes Week, in order to compete in the principal events.

Several centreboarders have been presented to members of the royal family by organizations or foreign countries: it is usual for such boats to be permanently loaned to schools or clubs, for the benefit of people who would not otherwise have the chance to sail.

The Royal Cutter, *Brittannia*, 122 ft, sailing in her last race in 1935. The rigging here is her final one which was under the J-class rules (compare with original rig on page 12).

Centre
From 1947 to 1965, the Queen and Duke of Edinburgh owned *Bluebottle*, which was raced regularly in the Dragon class. On board here are Prince Philip, the young Prince Charles and the late Uffa Fox (facing forward).

Right
Bloodhound, with Prince Philip at the helm. An ocean racer after the Dorade pattern, she was built before the Second World War, but extensively re-rigged under the direction of John Illingworth and sailed by the Royal Family in the 1960s.

The fastest building of any royally owned yacht was that of *White Rose*. She was a 1-rater class sloop of 23 ft designed and built by Charles Sibbick of Cowes in 1895. She was required by the Duke of York (later George V) to be ready for him when on leave from the Navy for racing at Cowes. She was built in five and a half days under the direction of the foreman of the yard, Ned Williams.

The busiest royal yacht was the *Royal Caroline*, built in 1749. She was ship rigged with length 90 ft, beam 25 ft, and tonnage 232. In effect she was a naval vessel and was first used for conveying King George II to and from the Continent, which included his frequent visits to Hanover. She entered the reign of George III and in 1761 was renamed *Royal Charlotte* when she sailed to Germany to fetch the King's bride, Princess Charlotte of Mecklenburg-Strelitz. The vessel was dismantled in 1820.

The last yacht of King George III was *Royal Sovereign*, 96 ft, ship rigged. Built at Deptford in 1805, she day sailed with the King on board off Weymouth in 1805. She took King Louis XVIII to France in 1814 after the abdication of Napoleon. After that she was used by the Navy and broken up in 1850. She was extravagantly decorated and the old carved quarter badges are preserved at Portsmouth.

The 1-rater, *White Rose*, owned by the Duke of York (later King George V). Here (to windward of *Whisper*) she races in light airs off Portsmouth Harbour.

The yacht of King George III, *Royal Sovereign*, 96 ft, seen in Weymouth Bay, a favourite spot for the King and now popular for national racing fixtures (Painting by J. T. Seres).

The first royal yacht to sail with the newly formed Yacht Club was the *Royal George* in 1821 when George IV had joined the club and taken a 'cottage' at Cowes. With the club yachts, the King sailed from Cowes towards the Needles and then returned to anchor off his cottage and went ashore. The yacht, another three-masted ship-rigged vessel, had been built in Deptford in 1817 and was length 103 ft, beam 26 ft 6 in. King George IV went to Scotland in her in 1822 and Queen Victoria did so in 1842. The Queen found the passage north too slow and returned by chartered steamer, thus making it the point at which royalty changed from sail to steam. The first royal steam-yacht was put in hand after this. *Royal George* ended her days as a barrack hulk and was broken up in 1905.

The most used name for royal yachts was *Victoria and Albert*: there were three successive yachts of this name, which is confusing. The first royal steam yacht was the *V & A* built after Queen Victoria's slow trip to Scotland in *Royal George*. Sir Robert Peel wrote to the Queen in September 1842: 'Reference to your majesty's note of yesterday . . . the first act on my return . . . was to write to Lord Haddington and strongly urge upon the Admiralty the necessity of providing a steam yacht for your majesty's accommodation.' The Victorians did not procrastinate. The new *V & A* was laid down in November 1842, launched in April the following year at

The second *Victoria and Albert* launched in 1855 and broken up in 1904.

Pembroke, Wales. She was wood-built with paddle-wheels, with one funnel. Length was 200 ft, beam 33 ft, displacement 1034 tons. In the yacht in 1843, the Queen visited Cowes where she saw Osborne House and decided to buy it, a decision which had a lasting effect on the growth of Cowes Week.

The second *V & A* was launched at Pembroke in 1855. She was also wood-built, with paddle-wheels, two funnels, and three masts. Length was 300 ft, beam 40 ft 3 in, displacement 2470 tons. Speed was $14\frac{3}{4}$ knots. The third *V & A* was built in 1899 and also of wood construction, but was driven by twin screws and was length 380 ft, beam 50 ft, displacement 5500 tons. Queen Victoria never sailed in her; she was used by Edward VII and George V. The three *V & A*s were broken up respectively in 1868, 1904, and 1954.

The most royal steam yachts in commission at any one time was not surprisingly at the height of British affluence in 1863. There were four. Besides the second *V & A*, there was *Osborne*, which was the old *V & A*, *Elfin* and *Alberta*, the last two being smaller vessels for short passages and river trips. The total number of royal yachts since the *Mary* of 1660 is 45, but, as explained, this includes naval vessels classified in this way and not all were for the use of the sovereign, but for instance for the Viceroy of Ireland: some suffered early conversion, because of impending war, to naval use.

The third *Victoria and Albert*, the steam yacht used by Edward VII and George V.

BOATING ACCIDENTS

The record of disasters to craft of all sizes before the advent of steam was appalling. For instance in 1872, a late stage for commercial sail, 2000 vessels were lost around the British Isles. In the 1870s there was an average loss of 6 British ships per day in the world. But such figures must be compared with hazards of life at the time on shore through disease and so on.

The greatest number of annual small-craft accidents today are in the USA. In 1973 the US Coast Guard reported 1754 deaths in 'boating accidents'. There were in the year 1599 persons injured and property worth $11 376 000 was damaged. Altogether 6738 vessels were involved in 5322 accidents. Between 1961 and 1973 fatalities increased by 44 per cent, but over the same time there was a decrease from 22·1 fatalities per 100 000 boats to 21·9.

The greatest number of rescue operations handled by Britain's HM Coastguard to date in any one year was in 1974. There were 5572 'operations', 268 lives were lost and 70 per cent of the activities were in the 'recreational class'. In 1769 incidents 'sailing and power craft' were involved; 1377 operations were searches for missing vessels or persons; 559 cases were with dinghies and inflatables. In Britain the role of the Coastguard is a co-ordinating one between the lifeboat service, RAF and Navy helicopters, other vessels, and its own shore-based rescue parties.

The fastest known rescue by an outside agency was on 22 September 1974, when a 22 ft cruiser with two men on board began to sink in the Solent. At 12.46 flares were sent up, seen from the shore, and one minute later contact had been made by a coastguard station with a naval helicopter. By 12.48 the men, Ian Hardy and Philip Burrow, had been winched into the helicopter in the charge of Lieutenant Roger Asbey, RN.

The biggest sailing yachts to be involved in a serious collision which resulted in one of them sinking were *Satanita*, 131 ft 6 in, owned by A. D. Clarke and the America's Cup challenger of Lord Dunraven *Valkyrie II*. *Satanita* had the longest waterline length of any sailing-cutter (3 ft longer than *Reliance*) at 93 ft 6 in. The collision was at the Clyde Regatta of 5 July 1894. *Satanita* in trying to avoid a lug-sail half-decker with four men on board, had to bear away but was unable to luff up in time to avoid *Valkyrie II*.

Valkyrie II with Lord Dunraven at the helm tried to luff, but *Satanita* smashed through her just abaft the mast. *Satanita* had a hole in her bows, 10 ft long. *Valkyrie II* was locked to her, but broke free, damaged two spectator steam yachts and sank nine minutes after the collision. One of her crew was killed. A court action resulted, which ended in the House of Lords and revealed an anomaly between the Merchant Shipping Act and rules of yacht racing.

Accidents today

The most common accident depends on the type of boat in use, as do the most likely consequences, for instance:

Commercial shipping: accidents are infrequent but when they occur, they are costly in comparison with small craft.

Small runabouts and outboards: running out of fuel, engine failure – or, fortunately rarely, injuring swimmers with propellers.

Sailing-dinghies: capsize and swamping.

Keeled yachts and motor cruisers: running aground, driven ashore, or catching fire.

Toy rubber rafts and airbeds: drifting out to sea when there is an offshore wind. It was reported in 1974 that this type of incident was becoming more prevalent.

The best prevention for all types of accident at sea is thorough preparation appropriate to the type of craft and passage to be undertaken and concentration on his job – though the trip may be for pleasure – by the person in charge of the boat.

The most boats to be affected by a single spell of bad weather were on Long Island Sound in 1957. In this usually light-weather area with hundreds of boats of all sizes out at a week-end, sudden gale force winds caused over 300 boats to be in difficulties.

The most widespread difficulties experienced by pleasure craft off the coast of Britain over a consecutive period were caused by the gales of 1 to 3 September 1974. On the south and east coasts of England 66 lifeboats were launched. There were about 67 craft or groups of small craft which needed or were thought to need help and 57 persons were rescued. In the period,

From left to right
The first Enterprise has capsized and then after rounding the mark the second one gets out of control trying to avoid her and capsizes to windward.

A sailing-dinghy capsize is usually part of the sport. Note the considerable buoyancy of this Laser which has not taken in any water.

An IOR boat gets a knockdown under spinnaker. She appears to have gybed so that the mainsail is also pinning her down. She will recover because of her deep keel.

The remains of *Morning Cloud* are brought ashore at Shoreham Harbour.

two bad weather systems, depressions, passed over the British Isles. These were connected to ex-tropical hurricanes which had come eastward across the Atlantic and the first winds reached 40 knots and in the second 60 knots. Incidents included crews rescued from anchored yachts later smashed on rocks; towing craft to safety; pulling out capsized dinghies; standing by while yachts made port; answering flares and signals.

The worst incident in this gale was when the expensively equipped ocean-racing yacht *Morning Cloud*, 43 ft, owned by the Rt Hon. Edward Heath, with a skipper and crew delivering her from Burnham-on-Crouch to Hamble ran into exceptional seas off the Sussex coast, east of the Owers Lightship. The yacht was hove on her beam ends and a man was lost overboard. After sailing back and failing to find him, the yacht resumed her course and probably ran into a similar patch of exceptional breaking sea. The yacht again was knocked down, undergoing severe structural damage and another member of the crew was swept overboard. Expecting the yacht to sink shortly, the remaining six men took to the four-man liferaft (the six-man one had been washed overboard when the yacht was flung down) which after eight hours was blown ashore with the six survivors on a beach. The yacht was later found on the seabed and the remains were salvaged.

The worst disasters in a single regular race of the Royal Ocean Racing Club were in the 1964 course from Santander, Spain to La Trinité, France across the Bay of Biscay. Five men died, from two yachts, all French, when a severe gale hit the fleet sailing northward. The owner of one yacht, *Aloa*, was swept overboard near the Rochbonne Bank (a dangerous shoal), and four of the crew of six of *Marie-Galante II* were lost when she foundered trying to run for shelter to the Gironde Estuary. Thirty-four other yachts, however, completed the course safely to La Trinité.

The most practical ways which you can take to achieve safety on the water are those concerned primarily with attitude and knowledge rather than technicalities. Technical integrity is important, but arises from attitudes of designer, builder, and sailor rather than being the result of a code or rules. In most countries there is increasing interest by the authorities in the use of small pleasure craft. Invariably there is either legislation of greater or lesser extent or at the least official recommendations. The situation in Britain is of the latter category. All who go afloat are urged to acquire knowledge and experience by reading, night classes, courses, and discussions as well as by practice under first easy conditions and then more difficult. However, the official aids and rules can be summarized under the basic design and construction of the boat, its equipment, and personal proficiency as follows:

Basic design

For craft under 20 ft there should be an approval plate affixed from the Ship and Boat Builders' National Federation. This does not mean that each individual boat has been inspected. For larger yachts and boats, the yard should hold a Lloyd's Certificate which implies that the conditions under which the boat was built are passed by Lloyd's, which looks at sample production boats and makes spot checks. This applies to GRP, but boats in other materials can be built singly under Lloyd's supervision.

Equipment

Racing boats, one-designs, and centreboarders must comply with class rules, which include minimum weights, specified construction and buoyancy tests for the boat. Offshore sailing boats should comply with the safety and special regulations of the Offshore Rating Council, which are international, reviewed annually, and may be amended or added to locally. Power-boats have to comply with class requirements which vary with size and depending whether they are offshore or inland.

For all craft in Britain under 45 ft there is a set of official recommendations of safety equipment. A supplementary list is for craft under 18 ft. These are issued by the Department of Trade.

There are British Standards issued by the British Standards Institution for life-jackets (this is most important as there are many variations sold as life-jackets), lifelines and harnesses, fire-extinguishers and inflatable boats.

Proficiency

Everyone is expected to know the International Regulations for Prevention of Collision at Sea – the Rules of the Road. They have the force of law in the event of accident. To them must be added local laws which may apply to harbours, rivers, and waterways. The international rule of the road is due to alter in 1976. It was agreed by all countries at the Law of the Sea Conference in 1974 but subsequently needs to be ratified by national parliaments.

In Britain there is now a system of voluntary proficiency certificates awarded by the Royal Yachting Association. These include day-boat certificates, coastal and offshore certificates, and power-boat certificates. Numerous sailing schools, of which the RYA keeps a register, run classes and courses to aid in the attainment of these standards. Such courses are ashore in the class-room and afloat, but need to be backed up by frequent practical experience.

Index